Solving equations
using modified Fibonacci
sequences
- an observation

November 2010
Peter Müller

Bibliografische Information Der Deutschen Bibliothek
Die Deutsche Bibliothek verzeichnet diese Publikation in der Deutschen
Nationalbibliografie; detaillierte bibliografische Daten sind im Internet über
http://dnb.ddb.de abrufbar

1. Auflage 2010.11.20, Version 2.2

Herstellung und Verlag: Books on Demand GmbH,
Norderstedt

© 2010 Peter Müller

e-mail: pmueller@gmxpro.de

ISBN: 9783842339620

it all began on top of the roof of the hotel in
Heiligendamm…thanks Angie

…and special thanks to Dr. Ruppert Gnatz, I would have
stopped before reaching the end.

The author:

Peter Müller was born in 1962 in Heidelberg. After graduating as electrical engineer in 1992 at the University of Applied Sciences in Mannheim he started to work in the field of functional safety of computerized systems.
Already during the years of study iterative processes like calculating the Mandelbrot set attracted the author's interest.
Since 2002 he has been associated with *exida.com*, a leading company for functional safety.

History

The Version 1 of this paper is based on the strong belief that solving mathematical problems has something to do with symmetry – I think the result shows this in a nice way.

The main thesis, shown in this paper, was developed by transferring rules which were found for simple examples to more complex cases – that's why the subtitle is "- an observation".

By the intensive help of Dr. Ruppert Gnatz it was possible to make the link to the "Power iteration" or "von Mises Iteration" to solve the problem. This leads to the published Version 2 of this paper.

Those who only want to know an easy approach to solving equations of higher degrees can skip chapter 1 of this second version – but then you miss the fun part of it.

Abstract

This paper is intended to report about the observation related to the possibility to use modified Fibonacci sequences, to solve equations of the form

$$x = 1 + b\frac{1}{x^{v-1}} \text{ , or } x^v - x^{v-1} - b = 0.$$

The method was developed in order to be able to determine at least one root for such equations v being higher than 4, as there is no easy way to solve equations with a higher degree than 4.

The mathematical proof of all thesis stated below is not done within this paper. They are only brought into the context of existing methods to solve equations of higher degrees.

The main thesis

For equations like

$$x^z - b_1 x^{z-1} - b_2 x^{z-2} - b_3 x^{z-3} - \ldots - b_z = 0$$

An approximation of a root can be found by calculating the ratio of two consecutive numbers $\lim\limits_{n \to \infty} \dfrac{a_n}{a_{n-1}}$ of a sequence that is generated by

$$a_n = b_1 a_{n-1} + b_2 a_{n-2} + b_3 a_{n-3} + \ldots + b_z a_{n-z}$$

if one number of $a_1 \ldots a_z \neq 0$ (and if the limes exist).

Assuming that the equation above can be solved by using a sequence of numbers the equation can be easily solved by using a table calculation program.
A more systematic way to do it is to use the following iterative matrix – vector multiplication:

The sequence of a_n can be reached by an iteration using the following quadratic matrix.

$$\vec{a}_i = \begin{pmatrix} 0 & 1 & 0 & \dots & 0 \\ 0 & 0 & 1 & \dots & 0 \\ \dots & \dots & \dots & \dots & \dots \\ 0 & 0 & 0 & \dots & 1 \\ b_z & b_{z-1} & b_{z-2} & \dots & b_1 \end{pmatrix} * \vec{a}_{i-1} = \begin{pmatrix} a_{i+1} \\ a_{i+2} \\ a_{i+3} \\ \dots \\ a_{i+z} \end{pmatrix}$$

whereby the start vector $\vec{a}_0 = \begin{pmatrix} 0 \\ 1 \\ 1 \\ \dots \\ 1 \end{pmatrix} = \begin{pmatrix} a_1 \\ a_2 \\ a_3 \\ \dots \\ a_z \end{pmatrix}$

Comparing this procedure with existing methods, the „von Mises iteration" (also called "Power iteration") can be found. This iteration calculates the normalized eigenvector of the matrix above. It can be shown that the greatest absolute eigenvalue of the matrix above is calculated by the ratio of two numbers of the eigenvector, or simply as the length of the (not normalized) eigenvector.
The eigenvalue is one of the values that solve the above equation.

Table of Content

1 The Observation

1.1 The Fibonacci sequence as basis

The Fibonacci numbers, which build the Fibonacci sequence, can be calculated by adding within a sequence of numbers the previous two numbers. The sequence starts per definition with 0 and 1.

a_1	a_2	a_3	a_4	a_5	a_6	a_7	a_8
0	1	1	2	3	5	8	13

a_9	a_{10}	a_{11}	a_{12}	a_{13}	a_{14}	a_{15}	a_n
21	34	55	89	144	233	377	...

Each number a_n is calculated by the relation:

$$a_n = a_{n-1} + a_{n-2} \qquad (1)$$

According to [D2] Johannes Kepler found that the ratio of two consecutive numbers converges to φ (the golden ratio: 1,6180339887…) (see Annex 1.1, b=1).

$$\lim_{n \to \infty} \frac{a_n}{a_{n-1}} = \varphi \qquad (2)$$

Using formula (1) by dividing it by a_{n-1}:

$$\frac{a_n}{a_{n-1}} = \frac{a_{n-1}}{a_{n-1}} + \frac{a_{n-2}}{a_{n-1}}$$

and assuming that every $\dfrac{a_m}{a_{m-1}}$ converges to x we get:

$$x = 1 + \frac{1}{x} \qquad\qquad (3)$$

Which equals: $x^2 - x - 1 = 0$

This is a quadratic equation having the general form

$$ax^2 + bx + c = 0$$

whereby the two roots can be calculated by the quadratic function[1]:

$$x_{1/2} = -\frac{b}{2a} \pm \sqrt{\frac{b^2}{4a^2} - c}$$

In (3) the factor a=1, b=-1 and c=-1, so x can be solved as

$$x_{1/2} = \frac{1}{2} \pm \sqrt{\frac{1}{4} + 1} = \frac{1 \pm \sqrt{5}}{2}$$

$$x_1 = \frac{1 + \sqrt{5}}{2} = 1,6180339887\ldots = \varphi$$

$$x_2 = \frac{1 - \sqrt{5}}{2} = -0,6180339887\ldots$$

Summary:

The ratio of two consecutive number of the Fibonacci sequence converges to $\lim\limits_{n \to \infty} \dfrac{a_n}{a_{n-1}} = \varphi$, which can be calculated by solving the equation $x = 1 + \dfrac{1}{x}$.

---

[1] As described in [D1] and [D2] for quadratic functions

1.2 A first modification of the Fibonacci sequence

To introduce a modified version of the Fibonacci sequence, the recurrence relation is changed to:

$$a_n = a_{n-1} + a_{n-3} \qquad (4)$$

The row starts with 0, 1 and 1[2]:

a_1	a_2	a_3	a_4	a_5	a_6	a_7	a_8
0	1	1	1	2	3	4	6

a_9	a_{10}	a_{11}	a_{12}	a_{13}	a_{14}	a_{15}	a_n
9	13	19	28	41	60	88	...

Like in chapter 1.1 the ratio of two consecutive numbers seems to converge to 1,465571 (see Annex 1.2, b=1). The same procedure to calculate the limit value is applied.

Using formula (4) by dividing it first by a_{n-1}:

$$\frac{a_n}{a_{n-1}} = \frac{a_{n-1}}{a_{n-1}} + \frac{a_{n-3}}{a_{n-1}} = \frac{a_{n-1} + a_{n-3}}{a_{n-1}}$$

and then extending the right part of the equation with the term a_{n-2}:

---

[2] Actually it seems not to be important which numbers to use, but it seems necessary that at least one number of a_1 to a_3 is different from 0.

$$\frac{a_n}{a_{n-1}} = \frac{a_{n-1} + a_{n-3}}{a_{n-1}} * \frac{a_{n-2}}{\dfrac{1}{a_{n-2}}} = \frac{\dfrac{1}{a_{n-2}}}{\dfrac{a_{n-1}}{a_{n-2}}} \cdot \frac{a_{n-1}}{a_{n-2}} + \frac{a_{n-3}}{a_{n-2}}$$

and assuming that every $\dfrac{a_m}{a_{m-1}}$ converges to x we get:

$$x = \frac{x + \dfrac{1}{x}}{x} = 1 + \frac{1}{x^2} \qquad (5)$$

Which equals: $x^3 - x^2 - 1 = 0$

This is a cubic equation which having the general form

$$ax^3 + bx^2 + cx + d = 0$$

whereby the roots of cubic equations can be determined in the following way[3]:

First dividing by a and then substituting $y = x + \dfrac{b}{3a}$ we get

$$y^3 + p^* y + q^* = 0$$

with

$$p^* = 3p = \frac{3ac - b^2}{3a^2} \quad \text{and} \quad q^* = 2q = \frac{2b^3 - 9abc + 27a^2 d}{27a^3}$$

---

[3] As described in [D1] and [D2] for cubic functions

Applying Cardano's Method:
The roots for y can be determined by setting $y = u + v$ where:

$$u = \sqrt[3]{-q + \sqrt{q^2 + p^3}} \text{ and } v = \sqrt[3]{-q - \sqrt{q^2 + p^3}}$$

if the condition $uv = -p$ is true.

In (5) the factor a=1, b=-1, c=0 and d=-1 so we get:

$$p = -\frac{1}{9} \qquad q = -\frac{29}{54}$$

$$y = \sqrt[3]{-q + \sqrt{q^2 + p^3}} + \sqrt[3]{-q - \sqrt{q^2 + p^3}}$$

This result in y=1,1322378…which after the back substitution equals x=1,465571…

Summary:
The ratio of two consecutive number of the modified Fibonacci sequence converges to $\lim\limits_{n \to \infty} \dfrac{a_n}{a_{n-1}}$, which can be calculated by solving the equation $x = 1 + \dfrac{1}{x^2}$.

1.3 A second modification of the Fibonacci sequence

To introduce a second modified version of the Fibonacci sequence, the recurrence relation is changed to:

$$a_n = a_{n-1} + a_{n-4} \qquad (6)$$

The row starts with 0, 1, 1 and 1:

a_1	a_2	a_3	a_4	a_5	a_6	a_7	a_8
0	1	1	1	1	2	3	4

a_9	a_{10}	a_{11}	a_{12}	a_{13}	a_{14}	a_{15}	a_n
5	7	10	14	19	26	36	...

Like in chapter 1.1 and 1.2 the ratio of two consecutive numbers seems to converge to 1,380278 (see Annex 1.3, b=1).
The same procedure to calculate the limit value is applied.

Using formula (6) by dividing it first by a_{n-1}:

$$\frac{a_n}{a_{n-1}} = \frac{a_{n-1}}{a_{n-1}} + \frac{a_{n-4}}{a_{n-1}} = \frac{a_{n-1} + a_{n-4}}{a_{n-1}}$$

and then extending the right part of the equation with the term a_{n-2}:

$$\frac{a_n}{a_{n-1}} = \frac{a_{n-1}+a_{n-4}}{a_{n-1}} * \frac{\dfrac{1}{a_{n-2}}}{\dfrac{1}{a_{n-2}}} = \frac{\dfrac{a_{n-1}}{a_{n-2}}+\dfrac{a_{n-4}}{a_{n-2}}}{\dfrac{a_{n-1}}{a_{n-2}}}$$

Furthermore the upper right part is extended by $\dfrac{a_{n-2}}{a_{n-3}}$

$$\frac{a_n}{a_{n-1}} = \frac{\dfrac{a_{n-1}}{a_{n-2}}+\dfrac{a_{n-4}}{a_{n-2}}*\dfrac{a_{n-2}}{a_{n-3}}\dfrac{a_{n-3}}{a_{n-2}}}{\dfrac{a_{n-1}}{a_{n-2}}} = \frac{\dfrac{a_{n-1}}{a_{n-2}}+\dfrac{a_{n-4}}{a_{n-3}}\dfrac{a_{n-3}}{a_{n-2}}}{\dfrac{a_{n-1}}{a_{n-2}}}$$

and assuming that every $\dfrac{a_m}{a_{m-1}}$ converges to x we get:

$$x = \frac{x + \dfrac{1}{x}\dfrac{1}{x}}{x} = 1 + \frac{1}{x^3} \tag{7}$$

Which equals: $x^4 - x^3 - 1 = 0$
This is a quartic function having the general form
$$ax^4 + bx^3 + cx^2 + dx + e = 0$$
whereby the roots can be calculated by using Lodovico Ferrari's Method[4]:

[4] As described in [D1] and [D2] for quartic functions

We reach the equation: $y^4 + \alpha y^2 + \beta y + \gamma = 0$ with:

$$y = x + \frac{b}{4a} \quad \text{and}$$

$$\alpha = -\frac{3b^2}{8a^2} + \frac{c}{a}, \qquad \beta = \frac{b^3}{8a^3} - \frac{bc}{2a^2} + \frac{d}{a},$$

$$\gamma = -\frac{3b^4}{256a^4} + \frac{b^2 c}{16a^3} - \frac{bd}{4a^2} + \frac{e}{a}$$

so, y is solved by

$$y = \frac{\pm_1 W \pm_2 \sqrt{-(\alpha + 2z) - 2\left(\alpha \pm_1 \dfrac{\beta}{W}\right)}}{2}$$

The two $\pm_1$ must have the same sign, the $\pm_2$ is independent.

where

$$W = \sqrt{\alpha + 2z}, \qquad z = -\frac{5}{6}\alpha - \frac{P}{3U} + U,$$

with

$$U = \sqrt[3]{-\frac{Q}{2} + \sqrt{\frac{Q^2}{4} + \frac{P^3}{27}}}, \qquad P = -\frac{\alpha^2}{12} - \gamma \quad \text{and}$$

$$Q = -\frac{\alpha^3}{108} + \frac{\alpha\gamma}{3} - \frac{\beta^2}{8}$$

In (7) the factor a=1, b=-1, c=0, d=0 and e=-1 so we get:

$$\alpha = -\frac{3}{8}; \qquad \beta = -\frac{1}{8}; \qquad \gamma = -\frac{3}{256} - 1; P=1; \quad Q=\frac{1}{8};$$

U=0,519057...; z=0,189366…; W=0,061105…

This result in y=1,130277…which after the back substitution equals x=1,380277…

Summary:
The ratio of two consecutive number of the modified Fibonacci sequence converges to $\displaystyle\lim_{n \to \infty} \frac{a_n}{a_{n-1}}$, which can be calculated by solving the equation $x = 1 + \dfrac{1}{x^3}$.

1.4 The first thesis

As shown in the chapters above, it seems that an approximation of one root of equations of the type

$$x = 1 + \frac{1}{x^{v-1}} \qquad (8)$$

can be found by calculating the ratio of two consecutive numbers $\displaystyle\lim_{n \to \infty} \frac{a_n}{a_{n-1}}$ of modified Fibonacci sequences, which are created by using the recurrence relation

$$a_n = a_{n-1} + a_{n-v} \qquad (9)$$

1.5 Another modification of the Fibonacci sequence

Another modification to the recurrence relation, used to calculate the Fibonacci sequence can be:

$$a_n = a_{n-1} + b * a_{n-2} \qquad (10)$$

Using formula (10) and dividing it by a_{n-1}

$$\frac{a_n}{a_{n-1}} = \frac{a_{n-1}}{a_{n-1}} + b \frac{a_{n-2}}{a_{n-1}}$$

and assuming that every $\dfrac{a_m}{a_{m-1}}$ converges to x we get:

$$x = 1 + b \frac{1}{x} \qquad (11)$$

Which is the same as: $x^2 - x - b = 0$ and can be solved in the identical way as shown in chapter 1.1.

1.6 The second thesis

It seems that an approximation for one root of equations of the type

$$x = 1 + b \frac{1}{x^{v-1}} \qquad (12)$$

can be found by calculating the ratio of two consecutive numbers $\lim\limits_{n \to \infty} \dfrac{a_n}{a_{n-1}}$ of modified Fibonacci sequences, which are created by using the recurrence relation

$$a_n = a_{n-1} + b * a_{n-v} \qquad (13)$$

1.7 A general approach to solve equations

Looking to the chapters before, it can be observed that there is a relation between an equation like $x^z - x^{z-1} - b = 0$ and an algorithm that generates sequences of numbers like $a_n = a_{n-1} + b * a_{n-z}$.

The relation is given, as the ratio of two consecutive numbers of the sequence can be used to calculate one of the roots of the equation.

This can be made more generic. The general form of such an algorithm

$$a_n = b_1 a_{n-1} + b_2 a_{n-2} + b_3 a_{n-3} + ... + b_z a_{n-z}$$

can be related to the general form of the equation

$$x^z - b_1 x^{z-1} - b_2 x^{z-2} - b_3 x^{z-3} - ... - b_z = 0$$

An approximation of a root of such an equation can be found by $\dfrac{a_n}{a_{n-1}}$.

A pre- condition for this is that $\dfrac{a_n}{a_{n-1}}$ converges:

$$\frac{a_n}{a_{n-1}} \approx \frac{a_{n-1}}{a_{n-2}}$$

1.8 The main thesis

For equations like

$$x^z - b_1 x^{z-1} - b_2 x^{z-2} - b_3 x^{z-3} - \ldots - b_z = 0 \qquad (14)$$

An approximation of a root can be found by calculating the ratio of two consecutive numbers $\lim\limits_{n \to \infty} \dfrac{a_n}{a_{n-1}}$ of a sequence that is generated by

$$a_n = b_1 a_{n-1} + b_2 a_{n-2} + b_3 a_{n-3} + \ldots + b_z a_{n-z} \qquad (15)$$

The first numbers of the sequence can be
$a_1 = 0; a_2 \ldots a_z = 1$

A condition is that the limit $\lim\limits_{n \to \infty} \dfrac{a_n}{a_{n-1}}$ exists.

1.9 Not a real proof – just a check

1.9.1 ...of first and the second thesis

The Annex 1 of this paper calculates the results for
$$\lim_{n \to \infty} \frac{a_n}{a_{n-1}}$$ for some different modified Fibonacci
sequences, in order to verify the correctness of the thesis
listed above – well, may be not really for $n \to \infty$, but n
was chosen to be 200.

It could be verified that for the following equations, with
different factors of b, one root could be found in that way
(see Annex 1.1 - 1.4):

	b=0,1	b=0,3	b=1	b=10	b=30
$x = 1 + b\dfrac{1}{x}$	x= 1,091607978	x= 1,24162	x= 1,618034	x= 3,701562	x= 6
$x = 1 + b\dfrac{1}{x^2}$	x= 1,084952904	x= 1,206198	x= 1,465571	x= 2,544512	x= 3,478849
$x = 1 + b\dfrac{1}{x^3}$	x= 1,079494742	x= 1,18177	x= 1,380278	x= 2,09209	x= 2,636661
$x = 1 + b\dfrac{1}{x^{10}}$	x= 1,057288283	x= 1,107786	x= 1,184276	x= 1,385008	x= 1,504925

For the grey shaded case n=200 is too small in order to
give a decent result.
The other cases show good results proving that the
method can be used to make an approximation of the
roots.

1.9.2 …of the main thesis

The Annex 2 of this paper calculates the results for

$$a_n = b_2 a_{n-2} + b_4 a_{n-4} + b_7$$

In order to find a root for the equation

$$x^7 - b_2 x^5 - b_4 x^3 - b_7 = 0$$

For $b_2 = 0,3; b_4 = 2,5; b_7 = 6$ we get

$$\frac{a_{200}}{a_{199}} = 1,49318428686666$$ which solves the equation
pretty good.

2 A more systematic approach

The general thesis was found by an observation as described before in this paper.
Here it will be shown that formula (14) above

$$x^z - b_1 x^{z-1} - b_2 x^{z-2} - b_3 x^{z-3} - \ldots - b_z = 0$$

can be transferred to formula (15) above

$$a_n = b_1 a_{n-1} + b_2 a_{n-2} + b_3 a_{n-3} + \ldots + b_z a_{n-z}$$

by the substitution $x^i \rightarrow \dfrac{a_{n+i}}{a_n}$:

$$\frac{a_{n+z}}{a_n} - b_1 \frac{a_{n+(z-1)}}{a_n} - b_2 \frac{a_{n+(z-2)}}{a_n} - b_3 \frac{a_{n+(z-3)}}{a_n} - \ldots - b_z = 0$$

$$a_{n+z} - b_1 a_{n+(z-1)} - b_2 a_{n+(z-2)} - b_3 a_{n+(z-3)} - \ldots - b_z a_n = 0$$

and shifting the sequence by "–z" we get:

$$a_n - b_1 a_{n-1} - b_2 a_{n-2} - b_3 a_{n-3} - \ldots - b_z a_{n-z} = 0$$

which is identical to formula (15). So the observation that was reached by looking at modified Fibonacci rows, can be confirmed, by a simple substitution.
An approximation of a root can be found by calculating the ratio of two consecutive numbers $\lim\limits_{n \to \infty} \dfrac{a_n}{a_{n-1}}$.

2.1 Bringing it into context

The sequence of a_n can be generated by using a simple table calculation program, or iterating the following matrix multiplication:

$$\begin{pmatrix} a_{i+1} \\ a_{i+2} \\ a_{i+3} \\ ... \\ a_{i+z} \end{pmatrix} = \begin{pmatrix} 0 & 1 & 0 & ... & 0 \\ 0 & 0 & 1 & ... & 0 \\ ... & ... & ... & ... & ... \\ 0 & 0 & 0 & ... & 1 \\ b_z & b_{z-1} & b_{z-2} & ... & b_1 \end{pmatrix} * \begin{pmatrix} a_{(i-1)+1} \\ a_{(i-1)+2} \\ a_{(i-1)+3} \\ ... \\ a_{(i-1)+z} \end{pmatrix}$$

For the first step the start vector $\vec{a}_0$ is:

$$\vec{a}_0 = \begin{pmatrix} 0 \\ 1 \\ 1 \\ ... \\ 1 \end{pmatrix} = \begin{pmatrix} a_1 \\ a_2 \\ a_3 \\ ... \\ a_z \end{pmatrix} \text{, it has the size of z.}$$

The following example shows this for the equation:

$$x^3 - x^2 - 0{,}64x + 0{,}512 = 0$$

whereby: $b_1 = 1$; $b_2 = 0{,}64$; $b_3 = -0{,}512$; $z = 3$

We get the matrix and the start vector:

$$A = \begin{pmatrix} 0 & 1 & 0 \\ 0 & 0 & 1 \\ -0,512 & 0,64 & 1 \end{pmatrix} \qquad \vec{a}_0 = \begin{pmatrix} 0 \\ 1 \\ 1 \end{pmatrix} = \begin{pmatrix} a_1 \\ a_2 \\ a_3 \end{pmatrix}$$

By iterating the multiplication we get $\vec{a}_i = A * \vec{a}_{i-1}$:

$$\vec{a}_1 = \begin{pmatrix} 1 \\ 1 \\ 1,64 \end{pmatrix} = \begin{pmatrix} a_2 \\ a_3 \\ a_4 \end{pmatrix}; \qquad \vec{a}_2 = \begin{pmatrix} 1 \\ 1,64 \\ 1,768 \end{pmatrix} = \begin{pmatrix} a_3 \\ a_4 \\ a_5 \end{pmatrix};$$

$$\vec{a}_3 = \begin{pmatrix} 1,64 \\ 1,768 \\ 2,306 \end{pmatrix} = \begin{pmatrix} a_4 \\ a_5 \\ a_6 \end{pmatrix}; \ldots \vec{a}_{40} = \begin{pmatrix} 618,829 \\ 726,257 \\ 852,334 \end{pmatrix} = \begin{pmatrix} a_{41} \\ a_{42} \\ a_{43} \end{pmatrix}$$

by doing the back-transformation we get

$$x \leftarrow \frac{a_n}{a_{n-1}} = \frac{a_{43}}{a_{42}} = 1,173599 \text{ which solves the above}$$

equation very well.

To keep the numbers in the vector $\vec{a}$ small the vector can be normalized before each multiplication. This means to divide each a_x by the length of the vector:

$$\vec{a}_{xN} \rightarrow \frac{1}{\|\vec{a}_x\|} * \vec{a}_x; \qquad \|\vec{a}\| = \sqrt{a_n^2 + a_{n-1}^2 + a_{n-2}^2} \ .$$

The start vector becomes: $\vec{a}_0 = \begin{pmatrix} 0 \\ 1 \\ 1 \end{pmatrix} \rightarrow \vec{a}_{0N} = \begin{pmatrix} 0 \\ 0,707 \\ 0,707 \end{pmatrix}$

$$\vec{a}_1 = \begin{pmatrix} 0,707 \\ 0,707 \\ 1,160 \end{pmatrix} \rightarrow \begin{pmatrix} 0,462 \\ 0,462 \\ 0,757 \end{pmatrix}; \qquad \vec{a}_2 = \begin{pmatrix} 0,462 \\ 0,757 \\ 0,816 \end{pmatrix} \rightarrow \begin{pmatrix} 0,383 \\ 0,628 \\ 0,677 \end{pmatrix};$$

$$\vec{a}_3 = \begin{pmatrix} 0,628 \\ 0,677 \\ 0,883 \end{pmatrix} \rightarrow \begin{pmatrix} 0,492 \\ 0,530 \\ 0,691 \end{pmatrix};$$

$$\vec{a}_{40} = \begin{pmatrix} 0,568 \\ 0,666 \\ 0,782 \end{pmatrix} \rightarrow \begin{pmatrix} 0,483686 \\ 0,567653 \\ 0,666197 \end{pmatrix}$$

by doing the back-transformation we get

$$x \leftarrow \frac{a_n}{a_{n-1}} = \frac{a_{43}}{a_{42}} = 1,173599 \text{ which solves the above}$$

equation as before.

2.2 Why does this work?

Actually this iteration of the matrix multiplication (using normalized vectors) is known as Power iteration or "von Mises iteration" (named after Richard von Mises).

According to [D2] the "von Mises iteration" is used to calculate the "eigenvector" of a matrix. By knowing the eigenvector of a matrix, the greatest absolute "eigenvalue" of the matrix can be found (by solving the "eigenvalue equation"). This is the value that solves the "characteristic equation" of the matrix.

But why does this solve our problem here? Let's take a look at it step by step:

First – the characteristic equation
We get the characteristic equation of a matrix by calculating the determinant $\det(A-\lambda E)$ and set it to 0, E being the identity matrix.
In our example above this means:

$$\det\left(\begin{pmatrix} 0 & 1 & 0 \\ 0 & 0 & 1 \\ -0{,}512 & 0{,}64 & 1 \end{pmatrix} - \begin{pmatrix} \lambda & 0 & 0 \\ 0 & \lambda & 0 \\ 0 & 0 & \lambda \end{pmatrix}\right) = 0$$

$$= \det\begin{pmatrix} -\lambda & 1 & 0 \\ 0 & -\lambda & 1 \\ -0{,}512 & 0{,}64 & 1-\lambda \end{pmatrix} = 0$$

Resulting in: $\lambda^3 - \lambda^2 - 0{,}64\lambda + 0{,}512 = 0$.
This is the identical equation we started with.

If the eigenvalue equation calculates the eigenvalue, which solves this equation, it becomes clear that our equation can be solved in that way.

Second – the eigenvector
The "von Mises iteration" results in the normalized Eigenvector $\vec{a}$, which is in our example above:

$$\vec{a}_{40N} = \begin{pmatrix} 0{,}483686 \\ 0{,}567653 \\ 0{,}666197 \end{pmatrix}$$

But why does the transformation $x \rightarrow \dfrac{a_n}{a_{n-1}}$ work?

Third – the eigenvalue
The so called "eigenvalue equation" combines the eigenvalue λ and the eigenvector $\vec{a}$ to the equation

$$(A-\lambda E)\ \vec{a}_{xN} = 0$$

In the example above we get:

$$\begin{pmatrix} -\lambda & 1 & 0 \\ 0 & -\lambda & 1 \\ -0{,}512 & 0{,}64 & 1-\lambda \end{pmatrix} * \begin{pmatrix} 0{,}483686 \\ 0{,}567653 \\ 0{,}666197 \end{pmatrix} = 0$$

This results in three formulas to calculate the eigenvalue:

$$-\lambda * 0{,}483686 + 0{,}567653 + 0 = 0$$
$$0 - \lambda * 0{,}567653 + 0{,}666197 = 0$$
$$-0{,}512 * 0{,}483686 + 0{,}64 * 0{,}567653 + (1 - \lambda)0{,}666197 = 0$$

All formulas resulting in the same value (1,173599) and the first two formulas result in

$$\lambda = \frac{0{,}567653}{0{,}483686} = \frac{a_{n-1}}{a_{n-2}} \text{ and } \lambda = \frac{0{,}666197}{0{,}567653} = \frac{a_n}{a_{n-1}}$$

This shows that the eigenvalue, which solves the characteristic equation (which is identical to the original equation to be solved) is calculated by the ratio of two consecutive numbers within the eigenvector.

As it is characteristic for the eigenvector that
- its direction is not changed by the multiplication with the matrix A and
- multiplication with matrix A or with the eigenvalue gives the identical resulting vector,

the eigenvalue can also be found as the length of the eigenvector before normalizing.

$$\vec{a}_{40} = \begin{pmatrix} 0{,}567653 \\ 0{,}666197 \\ 0{,}781847 \end{pmatrix};$$

$$\lambda = \|\vec{a}_{40}\| = \sqrt{0{,}567653^2 + 0{,}666197^2 + 0{,}781847^2} = 1{,}173599;$$

3 References

[D1]	Taschenbuch der Mathematik Bronstein / Semendjajew / Musiol / Mühlig, 7te Auflage, 2008
[D2]	The Free Encyclopedia "Wikipedia"

For further information please have a look in Wikipedia to:
1. Lineare Differenzengleichung (german)
2. Perron–Frobenius theorem

[D2] names those topics the following references:

1.1   B. Huppert: *Angewandte Lineare Algebra*, Walter de Gruyter (1990). ISBN 3-11-012107-7.

1.2   O. Perron: *Zur Theorie der Matrices*, Math. Ann. 64, 248-263 (1907).

1.3   G. Frobenius: *Über Matrizen aus nicht negativen Elementen*, Berl. Ber. 1912, 456-477.

2.1   Berg, L.: *Lineare Gleichungssysteme mit Bandstruktur*, München-Wien: Carl Hanser, 1986

2.2   Ian Jaques. *Mathematics for Economics and Business*, Fifth Edition. Prentice Hall, 2006. ISBN 0-273-70195-9. Chapter 9.1: Difference Equations, pp.551–568.

Annex 1 Example calculations

This Annex lists the calculation results for
$a_n = a_{n-1} + b * a_{n-v}$ used to determine roots for

$$x = 1 + b\frac{1}{x}, \quad x = 1 + b\frac{1}{x^2}, \quad x = 1 + b\frac{1}{x^3} \text{ and}$$

$x = 1 + b\dfrac{1}{x^{10}}$ in order to support the plausibility checks of

the first and the second thesis.

1.1 Calculation example for v-1=1

$\dfrac{a_{200}}{a_{199}}$

	1,091607978	1,24162	1,618034	3,701562	6
B	0,1	0,3	1	10	30
a_1	0	0	0	0	0
a_2	1	1	1	1	1
a_3	1	1	1	1	1
a_4	1,1	1,3	2	11	31
a_5	1,2	1,6	3	21	61
a_6	1,31	1,99	5	131	991
a_7	1,43	2,47	8	341	2821
a_8	1,561	3,067	13	1651	32551
a_9	1,704	3,808	21	5061	117181
a_{10}	1,8601	4,7281	34	21571	1093711
a_{11}	2,0305	5,8705	55	72181	4609141
a_{12}	2,21651	7,28893	89	287891	37420471
a_{13}	2,41956	9,05008	144	1009701	1,76E+08
a_{14}	2,641211	11,23676	233	3888611	1,3E+09
a_{15}	2,883167	13,95178	377	13985621	6,57E+09
a_{16}	3,1472881	17,32281	610	52871731	4,55E+10

a_{17}	3,4356048	21,50835	987	1,93E+08	2,43E+11
a_{18}	3,75033361	26,70519	1597	7,21E+08	1,61E+12
a_{19}	4,09389409	33,15769	2584	2,65E+09	8,89E+12
a_{20}	4,468927451	41,16925	4181	9,86E+09	5,71E+13
a_{21}	4,87831686	51,11656	6765	3,64E+10	3,24E+14
a_{22}	5,325209605	63,46733	10946	1,35E+11	2,04E+15
a_{23}	5,813041291	78,8023	17711	4,98E+11	1,17E+16
a_{24}	6,345562252	97,8425	28657	1,85E+12	7,29E+16
a_{25}	6,926866381	121,4832	46368	6,83E+12	4,25E+17
a_{26}	7,561422606	150,8359	75025	2,53E+13	2,61E+18
a_{27}	8,254109244	187,2809	121393	9,36E+13	1,54E+19
a_{28}	9,010251505	232,5317	196418	3,47E+14	9,37E+19
a_{29}	9,835662429	288,7159	317811	1,28E+15	5,55E+20
a_{30}	10,73668758	358,4754	514229	4,75E+15	3,37E+21
a_{31}	11,72025382	445,0902	832040	1,76E+16	2E+22
a_{32}	12,79392258	552,6329	1346269	6,51E+16	1,21E+23
a_{33}	13,96594796	686,1599	2178309	2,41E+17	7,21E+23
a_{34}	15,24534022	851,9498	3524578	8,92E+17	4,35E+24
a_{35}	16,64193502	1057,798	5702887	3,3E+18	2,6E+25
a_{36}	18,16646904	1313,383	9227465	1,22E+19	1,57E+26
a_{37}	19,83066254	1630,722	14930352	4,52E+19	9,36E+26
a_{38}	21,64730944	2024,737	24157817	1,67E+20	5,63E+27
a_{39}	23,6303757	2513,953	39088169	6,2E+20	3,37E+28
a_{40}	25,79510664	3121,375	63245986	2,29E+21	2,03E+29
a_{41}	28,15814421	3875,561	1,02E+08	8,49E+21	1,21E+30
a_{42}	30,73765488	4811,973	1,66E+08	3,14E+22	7,3E+30
a_{43}	33,5534693	5974,641	2,68E+08	1,16E+23	4,37E+31
a_{44}	36,62723479	7418,233	4,33E+08	4,31E+23	2,63E+32
a_{45}	39,98258172	9210,625	7,01E+08	1,59E+24	1,57E+33
a_{46}	43,64530519	11436,1	1,13E+09	5,9E+24	9,45E+33
a_{47}	47,64356337	14199,28	1,84E+09	2,18E+25	5,67E+34
a_{48}	52,00809389	17630,11	2,97E+09	8,09E+25	3,4E+35
a_{49}	56,77245022	21889,9	4,81E+09	2,99E+26	2,04E+36
a_{50}	61,97325961	27178,93	7,78E+09	1,11E+27	1,22E+37
a_{51}	67,65050463	33745,9	1,26E+10	4,1E+27	7,35E+37
a_{52}	73,84783059	41899,58	2,04E+10	1,52E+28	4,41E+38

a_{53}	80,61288106	52023,35	3,3E+10	5,62E+28	2,65E+39
a_{54}	87,99766412	64593,22	5,33E+10	2,08E+29	1,59E+40
a_{55}	96,05895222	80200,22	8,63E+10	7,7E+29	9,52E+40
a_{56}	104,8587186	99578,19	1,4E+11	2,85E+30	5,71E+41
a_{57}	114,4646139	123638,3	2,26E+11	1,05E+31	3,43E+42
a_{58}	124,9504857	153511,7	3,65E+11	3,9E+31	2,06E+43
a_{59}	136,3969471	190603,2	5,91E+11	1,45E+32	1,23E+44
a_{60}	148,8919957	236656,7	9,57E+11	5,35E+32	7,41E+44
a_{61}	162,5316904	293837,7	1,55E+12	1,98E+33	4,44E+45
a_{62}	177,42089	364834,7	2,5E+12	7,33E+33	2,67E+46
a_{63}	193,674059	452986	4,05E+12	2,71E+34	1,6E+47
a_{64}	211,416148	562436,4	6,56E+12	1E+35	9,6E+47
a_{65}	230,7835539	698332,2	1,06E+13	3,72E+35	5,76E+48
a_{66}	251,9251687	867063,1	1,72E+13	1,38E+36	3,45E+49
a_{67}	275,0035241	1076563	2,78E+13	5,09E+36	2,07E+50
a_{68}	300,1960409	1336682	4,49E+13	1,89E+37	1,24E+51
a_{69}	327,6963934	1659650	7,27E+13	6,98E+37	7,46E+51
a_{70}	357,7159974	2060655	1,18E+14	2,58E+38	4,48E+52
a_{71}	390,4856368	2558550	1,9E+14	9,56E+38	2,69E+53
a_{72}	426,2572365	3176747	3,08E+14	3,54E+39	1,61E+54
a_{73}	465,3058002	3944312	4,98E+14	1,31E+40	9,67E+54
a_{74}	507,9315239	4897336	8,07E+14	4,85E+40	5,8E+55
a_{75}	554,4621039	6080629	1,3E+15	1,8E+41	3,48E+56
a_{76}	605,2552563	7549830	2,11E+15	6,64E+41	2,09E+57
a_{77}	660,7014667	9374018	3,42E+15	2,46E+42	1,25E+58
a_{78}	721,2269923	11638967	5,53E+15	9,1E+42	7,52E+58
a_{79}	787,2971389	14451173	8,94E+15	3,37E+43	4,51E+59
a_{80}	859,4198382	17942863	1,45E+16	1,25E+44	2,71E+60
a_{81}	938,1495521	22278215	2,34E+16	4,62E+44	1,62E+61
a_{82}	1024,091536	27661074	3,79E+16	1,71E+45	9,75E+61
a_{83}	1117,906491	34344539	6,13E+16	6,33E+45	5,85E+62
a_{84}	1220,315645	42642861	9,92E+16	2,34E+46	3,51E+63
a_{85}	1332,106294	52946222	1,61E+17	8,67E+46	2,11E+64
a_{86}	1454,137858	65739081	2,6E+17	3,21E+47	1,26E+65
a_{87}	1587,348488	81622947	4,2E+17	1,19E+48	7,58E+65
a_{88}	1732,762273	1,01E+08	6,8E+17	4,4E+48	4,55E+66

a_{89}	1891,497122	1,26E+08	1,1E+18	1,63E+49	2,73E+67
a_{90}	2064,77335	1,56E+08	1,78E+18	6,02E+49	1,64E+68
a_{91}	2253,923062	1,94E+08	2,88E+18	2,23E+50	9,82E+68
a_{92}	2460,400397	2,41E+08	4,66E+18	8,25E+50	5,89E+69
a_{93}	2685,792703	2,99E+08	7,54E+18	3,06E+51	3,54E+70
a_{94}	2931,832743	3,71E+08	1,22E+19	1,13E+52	2,12E+71
a_{95}	3200,412013	4,61E+08	1,97E+19	4,19E+52	1,27E+72
a_{96}	3493,595287	5,72E+08	3,19E+19	1,55E+53	7,64E+72
a_{97}	3813,636488	7,11E+08	5,17E+19	5,74E+53	4,58E+73
a_{98}	4162,996017	8,82E+08	8,36E+19	2,12E+54	2,75E+74
a_{99}	4544,359666	1,1E+09	1,35E+20	7,86E+54	1,65E+75
a_{100}	4960,659268	1,36E+09	2,19E+20	2,91E+55	9,9E+75
a_{101}	5415,095234	1,69E+09	3,54E+20	1,08E+56	5,94E+76
a_{102}	5911,161161	2,1E+09	5,73E+20	3,99E+56	3,56E+77
a_{103}	6452,670685	2,6E+09	9,27E+20	1,48E+57	2,14E+78
a_{104}	7043,786801	3,23E+09	1,5E+21	5,46E+57	1,28E+79
a_{105}	7689,053869	4,01E+09	2,43E+21	2,02E+58	7,7E+79
a_{106}	8393,432549	4,98E+09	3,93E+21	7,48E+58	4,62E+80
a_{107}	9162,337936	6,19E+09	6,36E+21	2,77E+59	2,77E+81
a_{108}	10001,68119	7,68E+09	1,03E+22	1,03E+60	1,66E+82
a_{109}	10917,91498	9,54E+09	1,66E+22	3,79E+60	9,98E+82
a_{110}	11918,0831	1,18E+10	2,69E+22	1,4E+61	5,99E+83
a_{111}	13009,8746	1,47E+10	4,36E+22	5,2E+61	3,59E+84
a_{112}	14201,68291	1,83E+10	7,05E+22	1,92E+62	2,15E+85
a_{113}	15502,67037	2,27E+10	1,14E+23	7,12E+62	1,29E+86
a_{114}	16922,83866	2,82E+10	1,85E+23	2,64E+63	7,76E+86
a_{115}	18473,1057	3,5E+10	2,99E+23	9,76E+63	4,65E+87
a_{116}	20165,38957	4,34E+10	4,83E+23	3,61E+64	2,79E+88
a_{117}	22012,70014	5,39E+10	7,82E+23	1,34E+65	1,68E+89
a_{118}	24029,23909	6,69E+10	1,26E+24	4,95E+65	1,01E+90
a_{119}	26230,50911	8,31E+10	2,05E+24	1,83E+66	6,03E+90
a_{120}	28633,43302	1,03E+11	3,31E+24	6,78E+66	3,62E+91
a_{121}	31256,48393	1,28E+11	5,36E+24	2,51E+67	2,17E+92
a_{122}	34119,82723	1,59E+11	8,67E+24	9,29E+67	1,3E+93
a_{123}	37245,47562	1,97E+11	1,4E+25	3,44E+68	7,82E+93
a_{124}	40657,45835	2,45E+11	2,27E+25	1,27E+69	4,69E+94

a_{125}	44382,00591	3,04E+11	3,67E+25	4,71E+69	2,81E+95
a_{126}	48447,75174	3,78E+11	5,94E+25	1,74E+70	1,69E+96
a_{127}	52885,95233	4,69E+11	9,62E+25	6,46E+70	1,01E+97
a_{128}	57730,72751	5,83E+11	1,56E+26	2,39E+71	6,08E+97
a_{129}	63019,32274	7,23E+11	2,52E+26	8,85E+71	3,65E+98
a_{130}	68792,39549	8,98E+11	4,07E+26	3,28E+72	2,2E+99
a_{131}	75094,32777	1,12E+12	6,59E+26	1,21E+73	1,3E+100
a_{132}	81973,56732	1,38E+12	1,07E+27	4,49E+73	7,9E+100
a_{133}	89483,00009	1,72E+12	1,73E+27	1,66E+74	4,7E+101
a_{134}	97680,35682	2,13E+12	2,79E+27	6,15E+74	2,8E+102
a_{135}	106628,6568	2,65E+12	4,52E+27	2,28E+75	1,7E+103
a_{136}	116396,6925	3,29E+12	7,31E+27	8,43E+75	1E+104
a_{137}	127059,5582	4,09E+12	1,18E+28	3,12E+76	6,1E+104
a_{138}	138699,2275	5,07E+12	1,91E+28	1,15E+77	3,7E+105
a_{139}	151405,1833	6,3E+12	3,1E+28	4,27E+77	2,2E+106
a_{140}	165275,106	7,82E+12	5,01E+28	1,58E+78	1,3E+107
a_{141}	180415,6243	9,71E+12	8,11E+28	5,85E+78	7,9E+107
a_{142}	196943,1349	1,21E+13	1,31E+29	2,17E+79	4,8E+108
a_{143}	214984,6974	1,5E+13	2,12E+29	8,02E+79	2,9E+109
a_{144}	234679,0109	1,86E+13	3,43E+29	2,97E+80	1,7E+110
a_{145}	256177,4806	2,31E+13	5,56E+29	1,1E+81	1E+111
a_{146}	279645,3817	2,87E+13	8,99E+29	4,07E+81	6,2E+111
a_{147}	305263,1298	3,56E+13	1,45E+30	1,51E+82	3,7E+112
a_{148}	333227,6679	4,42E+13	2,35E+30	5,57E+82	2,2E+113
a_{149}	363753,9809	5,48E+13	3,81E+30	2,06E+83	1,3E+114
a_{150}	397076,7477	6,81E+13	6,16E+30	7,64E+83	8E+114
a_{151}	433452,1458	8,45E+13	9,97E+30	2,83E+84	4,8E+115
a_{152}	473159,8206	1,05E+14	1,61E+31	1,05E+85	2,9E+116
a_{153}	516505,0351	1,3E+14	2,61E+31	3,87E+85	1,7E+117
a_{154}	563821,0172	1,62E+14	4,22E+31	1,43E+86	1E+118
a_{155}	615471,5207	2,01E+14	6,83E+31	5,31E+86	6,2E+118
a_{156}	671853,6224	2,49E+14	1,11E+32	1,96E+87	3,7E+119
a_{157}	733400,7745	3,1E+14	1,79E+32	7,27E+87	2,2E+120
a_{158}	800586,1367	3,85E+14	2,89E+32	2,69E+88	1,3E+121
a_{159}	873926,2142	4,78E+14	4,68E+32	9,96E+88	8,1E+121
a_{160}	953984,8279	5,93E+14	7,58E+32	3,69E+89	4,8E+122

a_{161}	1041377,449	7,36E+14	1,23E+33	1,37E+90	2,9E+123
a_{162}	1136775,932	9,14E+14	1,98E+33	5,05E+90	1,7E+124
a_{163}	1240913,677	1,13E+15	3,21E+33	1,87E+91	1E+125
a_{164}	1354591,27	1,41E+15	5,19E+33	6,92E+91	6,3E+125
a_{165}	1478682,638	1,75E+15	8,4E+33	2,56E+92	3,8E+126
a_{166}	1614141,765	2,17E+15	1,36E+34	9,49E+92	2,3E+127
a_{167}	1762010,029	2,7E+15	2,2E+34	3,51E+93	1,4E+128
a_{168}	1923424,205	3,35E+15	3,56E+34	1,3E+94	8,1E+128
a_{169}	2099625,208	4,16E+15	5,76E+34	4,81E+94	4,9E+129
a_{170}	2291967,629	5,16E+15	9,32E+34	1,78E+95	2,9E+130
a_{171}	2501930,149	6,41E+15	1,51E+35	6,59E+95	1,8E+131
a_{172}	2731126,912	7,96E+15	2,44E+35	2,44E+96	1,1E+132
a_{173}	2981319,927	9,88E+15	3,95E+35	9,03E+96	6,3E+132
a_{174}	3254432,618	1,23E+16	6,39E+35	3,34E+97	3,8E+133
a_{175}	3552564,611	1,52E+16	1,03E+36	1,24E+98	2,3E+134
a_{176}	3878007,873	1,89E+16	1,67E+36	4,58E+98	1,4E+135
a_{177}	4233264,334	2,35E+16	2,71E+36	1,7E+99	8,2E+135
a_{178}	4621065,121	2,92E+16	4,38E+36	6,3E+99	4,9E+136
a_{179}	5044391,555	3,62E+16	7,08E+36	2,3E+100	2,9E+137
a_{180}	5506498,067	4,5E+16	1,15E+37	8,6E+100	1,8E+138
a_{181}	6010937,222	5,58E+16	1,85E+37	3,2E+101	1,1E+139
a_{182}	6561587,029	6,93E+16	3E+37	1,2E+102	6,4E+139
a_{183}	7162680,751	8,6E+16	4,86E+37	4,4E+102	3,8E+140
a_{184}	7818839,454	1,07E+17	7,86E+37	1,6E+103	2,3E+141
a_{185}	8535107,529	1,33E+17	1,27E+38	6E+103	1,4E+142
a_{186}	9316991,475	1,65E+17	2,06E+38	2,2E+104	8,3E+142
a_{187}	10170502,23	2,04E+17	3,33E+38	8,2E+104	5E+143
a_{188}	11102201,38	2,54E+17	5,39E+38	3E+105	3E+144
a_{189}	12119251,6	3,15E+17	8,71E+38	1,1E+106	1,8E+145
a_{190}	13229471,74	3,91E+17	1,41E+39	4,2E+106	1,1E+146
a_{191}	14441396,9	4,86E+17	2,28E+39	1,5E+107	6,4E+146
a_{192}	15764344,07	6,03E+17	3,69E+39	5,7E+107	3,9E+147
a_{193}	17208483,76	7,49E+17	5,97E+39	2,1E+108	2,3E+148
a_{194}	18784918,17	9,3E+17	9,66E+39	7,8E+108	1,4E+149
a_{195}	20505766,54	1,16E+18	1,56E+40	2,9E+109	8,3E+149
a_{196}	22384258,36	1,43E+18	2,53E+40	1,1E+110	5E+150

Solving equations - using modified Fibonacci sequences
- an observation

a_{197}	24434835,01	1,78E+18	4,09E+40	4E+110	3E+151
a_{198}	26673260,85	2,21E+18	6,62E+40	1,5E+111	1,8E+152
a_{199}	29116744,35	2,74E+18	1,07E+41	5,4E+111	1,1E+153
a_{200}	31784070,43	3,41E+18	1,73E+41	2E+112	6,5E+153

1.2 Calculation example for v-1=2

$\dfrac{a_{200}}{a_{199}}$	1,084952904	1,206198	1,465571	2,544512	3,478849
b	0,1	0,3	1	10	30
a_1	0	0	0	0	0
a_2	1	1	1	1	1
a_3	1	1	1	1	1
a_4	1	1	1	1	1
a_5	1,1	1,3	2	11	31
a_6	1,2	1,6	3	21	61
a_7	1,3	1,9	4	31	91
a_8	1,41	2,29	6	141	1021
a_9	1,53	2,77	9	351	2851
a_{10}	1,66	3,34	13	661	5581
a_{11}	1,801	4,027	19	2071	36211
a_{12}	1,954	4,858	28	5581	121741
a_{13}	2,12	5,86	41	12191	289171
a_{14}	2,3001	7,0681	60	32901	1375501
a_{15}	2,4955	8,5255	88	88711	5027731
a_{16}	2,7075	10,2835	129	210621	13702861
a_{17}	2,93751	12,40393	189	539631	54967891
a_{18}	3,18706	14,96158	277	1426741	2,06E+08
a_{19}	3,45781	18,04663	406	3532951	6,17E+08
a_{20}	3,751561	21,76781	595	8929261	2,27E+09
a_{21}	4,070267	26,25628	872	23196671	8,44E+09
a_{22}	4,416048	31,67027	1278	58526181	2,69E+10
a_{23}	4,7912041	38,20061	1873	1,48E+08	9,49E+10
a_{24}	5,1982308	46,0775	2745	3,8E+08	3,48E+11

a_{25}	5,6398356	55,57858	4023	9,65E+08	1,16E+12
a_{26}	6,11895601	67,03877	5896	2,44E+09	4E+12
a_{27}	6,63877909	80,86202	8641	6,24E+09	1,44E+13
a_{28}	7,20276265	97,53559	12664	1,59E+10	4,91E+13
a_{29}	7,814658251	117,6472	18560	4,03E+10	1,69E+14
a_{30}	8,47853616	141,9058	27201	1,03E+11	6,03E+14
a_{31}	9,198812425	171,1665	39865	2,62E+11	2,08E+15
a_{32}	9,98027825	206,4607	58425	6,65E+11	7,16E+15
a_{33}	10,82813187	249,0324	85626	1,69E+12	2,52E+16
a_{34}	11,74801311	300,3824	125491	4,31E+12	8,75E+16
a_{35}	12,74604093	362,3206	183916	1,1E+13	3,02E+17
a_{36}	13,82885412	437,0303	269542	2,79E+13	1,06E+18
a_{37}	15,00365543	527,145	395033	7,1E+13	3,69E+18
a_{38}	16,27825952	635,8412	578949	1,81E+14	1,28E+19
a_{39}	17,66114494	766,9503	848491	4,59E+14	4,45E+19
a_{40}	19,16151048	925,0938	1243524	1,17E+15	1,55E+20
a_{41}	20,78933643	1115,846	1822473	2,97E+15	5,38E+20
a_{42}	22,55545093	1345,931	2670964	7,57E+15	1,87E+21
a_{43}	24,47160197	1623,459	3914488	1,93E+16	6,53E+21
a_{44}	26,55053562	1958,213	5736961	4,9E+16	2,27E+22
a_{45}	28,80608071	2361,992	8407925	1,25E+17	7,89E+22
a_{46}	31,25324091	2849,03	12322413	3,17E+17	2,75E+23
a_{47}	33,90829447	3436,494	18059374	8,07E+17	9,54E+23
a_{48}	36,78890254	4145,092	26467299	2,05E+18	3,32E+24
a_{49}	39,91422663	4999,801	38789712	5,23E+18	1,16E+25
a_{50}	43,30505608	6030,749	56849086	1,33E+19	4,02E+25
a_{51}	46,98394633	7274,277	83316385	3,38E+19	1,4E+26
a_{52}	50,97536899	8774,217	1,22E+08	8,61E+19	4,87E+26
a_{53}	55,3058746	10583,44	1,79E+08	2,19E+20	1,69E+27
a_{54}	60,00426923	12765,73	2,62E+08	5,58E+20	5,89E+27
a_{55}	65,10180613	15397,99	3,84E+08	1,42E+21	2,05E+28
a_{56}	70,63239359	18573,02	5,63E+08	3,61E+21	7,13E+28
a_{57}	76,63282052	22402,74	8,26E+08	9,18E+21	2,48E+29
a_{58}	83,14300113	27022,14	1,21E+09	2,34E+22	8,62E+29
a_{59}	90,20624049	32594,04	1,77E+09	5,95E+22	3E+30
a_{60}	97,86952254	39314,87	2,6E+09	1,51E+23	1,04E+31

a_{61}	106,1838227	47421,51	3,81E+09	3,85E+23	3,63E+31
a_{62}	115,2044467	57199,72	5,58E+09	9,8E+23	1,26E+32
a_{63}	124,991399	68994,18	8,18E+09	2,49E+24	4,39E+32
a_{64}	135,6097812	83220,63	1,2E+10	6,34E+24	1,53E+33
a_{65}	147,1302259	100380,5	1,76E+10	1,61E+25	5,32E+33
a_{66}	159,6293658	121078,8	2,58E+10	4,11E+25	1,85E+34
a_{67}	173,1903439	146045	3,77E+10	1,04E+26	6,44E+34
a_{68}	187,9033665	176159,2	5,53E+10	2,66E+26	2,24E+35
a_{69}	203,8663031	212482,8	8,11E+10	6,77E+26	7,79E+35
a_{70}	221,1853375	256296,3	1,19E+11	1,72E+27	2,71E+36
a_{71}	239,9756741	309144	1,74E+11	4,38E+27	9,43E+36
a_{72}	260,3623044	372888,9	2,55E+11	1,11E+28	3,28E+37
a_{73}	282,4808382	449777,8	3,74E+11	2,84E+28	1,14E+38
a_{74}	306,4784056	542521	5,48E+11	7,22E+28	3,97E+38
a_{75}	332,514636	654387,7	8,03E+11	1,84E+29	1,38E+39
a_{76}	360,7627199	789321	1,18E+12	4,67E+29	4,8E+39
a_{77}	391,4105604	952077,3	1,73E+12	1,19E+30	1,67E+40
a_{78}	424,662024	1148394	2,53E+12	3,03E+30	5,81E+40
a_{79}	460,738296	1385190	3,71E+12	7,7E+30	2,02E+41
a_{80}	499,879352	1670813	5,43E+12	1,96E+31	7,03E+41
a_{81}	542,3455544	2015331	7,96E+12	4,98E+31	2,45E+42
a_{82}	588,419384	2430888	1,17E+13	1,27E+32	8,51E+42
a_{83}	638,4073192	2932132	1,71E+13	3,23E+32	2,96E+43
a_{84}	692,6418747	3536731	2,51E+13	8,21E+32	1,03E+44
a_{85}	751,4838131	4265998	3,67E+13	2,09E+33	3,58E+44
a_{86}	815,324545	5145637	5,38E+13	5,32E+33	1,25E+45
a_{87}	884,5887325	6206657	7,89E+13	1,35E+34	4,34E+45
a_{88}	959,7371138	7486456	1,16E+14	3,44E+34	1,51E+46
a_{89}	1041,269568	9030147	1,69E+14	8,76E+34	5,25E+46
a_{90}	1129,728442	10892144	2,48E+14	2,23E+35	1,83E+47
a_{91}	1225,702153	13138081	3,64E+14	5,67E+35	6,35E+47
a_{92}	1329,82911	15847125	5,33E+14	1,44E+36	2,21E+48
a_{93}	1442,801954	19114769	7,82E+14	3,67E+36	7,69E+48
a_{94}	1565,372169	23056193	1,15E+15	9,34E+36	2,68E+49
a_{95}	1698,35508	27810331	1,68E+15	2,38E+37	9,31E+49
a_{96}	1842,635276	33544761	2,46E+15	6,05E+37	3,24E+50

a_{97}	1999,172492	40461619	3,61E+15	1,54E+38	1,13E+51
a_{98}	2169,008001	48804719	5,29E+15	3,92E+38	3,92E+51
a_{99}	2353,271528	58868147	7,75E+15	9,96E+38	1,36E+52
a_{100}	2553,188777	71006633	1,14E+16	2,54E+39	4,74E+52
a_{101}	2770,089577	85648048	1,66E+16	6,45E+39	1,65E+53
a_{102}	3005,41673	1,03E+08	2,44E+16	1,64E+40	5,74E+53
a_{103}	3260,735608	1,25E+08	3,57E+16	4,18E+40	2E+54
a_{104}	3537,744566	1,5E+08	5,24E+16	1,06E+41	6,95E+54
a_{105}	3838,286239	1,81E+08	7,68E+16	2,7E+41	2,42E+55
a_{106}	4164,359799	2,19E+08	1,12E+17	6,88E+41	8,41E+55
a_{107}	4518,134256	2,64E+08	1,65E+17	1,75E+42	2,92E+56
a_{108}	4901,96288	3,18E+08	2,42E+17	4,46E+42	1,02E+57
a_{109}	5318,39886	3,84E+08	3,54E+17	1,13E+43	3,54E+57
a_{110}	5770,212285	4,63E+08	5,19E+17	2,88E+43	1,23E+58
a_{111}	6260,408573	5,58E+08	7,61E+17	7,34E+43	4,28E+58
a_{112}	6792,248459	6,73E+08	1,11E+18	1,87E+44	1,49E+59
a_{113}	7369,269688	8,12E+08	1,63E+18	4,75E+44	5,18E+59
a_{114}	7995,310545	9,8E+08	2,39E+18	1,21E+45	1,8E+60
a_{115}	8674,535391	1,18E+09	3,51E+18	3,08E+45	6,27E+60
a_{116}	9411,46236	1,43E+09	5,14E+18	7,83E+45	2,18E+61
a_{117}	10210,99341	1,72E+09	7,54E+18	1,99E+46	7,59E+61
a_{118}	11078,44695	2,07E+09	1,1E+19	5,07E+46	2,64E+62
a_{119}	12019,59319	2,5E+09	1,62E+19	1,29E+47	9,19E+62
a_{120}	13040,69253	3,02E+09	2,37E+19	3,28E+47	3,2E+63
a_{121}	14148,53723	3,64E+09	3,48E+19	8,35E+47	1,11E+64
a_{122}	15350,49655	4,39E+09	5,1E+19	2,12E+48	3,87E+64
a_{123}	16654,5658	5,3E+09	7,47E+19	5,41E+48	1,35E+65
a_{124}	18069,41952	6,39E+09	1,09E+20	1,38E+49	4,68E+65
a_{125}	19604,46918	7,7E+09	1,6E+20	3,5E+49	1,63E+66
a_{126}	21269,92576	9,29E+09	2,35E+20	8,91E+49	5,67E+66
a_{127}	23076,86771	1,12E+10	3,45E+20	2,27E+50	1,97E+67
a_{128}	25037,31463	1,35E+10	5,05E+20	5,77E+50	6,86E+67
a_{129}	27164,3072	1,63E+10	7,4E+20	1,47E+51	2,39E+68
a_{130}	29471,99397	1,97E+10	1,08E+21	3,73E+51	8,3E+68
a_{131}	31975,72543	2,37E+10	1,59E+21	9,5E+51	2,89E+69
a_{132}	34692,15615	2,86E+10	2,33E+21	2,42E+52	1E+70

Solving equations - using modified Fibonacci sequences   41
- an observation

a_{133}	37639,35555	3,45E+10	3,41E+21	6,15E+52	3,49E+70
a_{134}	40836,92809	4,16E+10	5E+21	1,57E+53	1,22E+71
a_{135}	44306,14371	5,02E+10	7,33E+21	3,98E+53	4,23E+71
a_{136}	48070,07927	6,06E+10	1,07E+22	1,01E+54	1,47E+72
a_{137}	52153,77207	7,31E+10	1,58E+22	2,58E+54	5,12E+72
a_{138}	56584,38645	8,81E+10	2,31E+22	6,56E+54	1,78E+73
a_{139}	61391,39437	1,06E+11	3,38E+22	1,67E+55	6,19E+73
a_{140}	66606,77158	1,28E+11	4,96E+22	4,25E+55	2,15E+74
a_{141}	72265,21022	1,55E+11	7,27E+22	1,08E+56	7,5E+74
a_{142}	78404,34966	1,87E+11	1,07E+23	2,75E+56	2,61E+75
a_{143}	85065,02682	2,25E+11	1,56E+23	7E+56	9,07E+75
a_{144}	92291,54784	2,71E+11	2,29E+23	1,78E+57	3,16E+76
a_{145}	100131,9828	3,27E+11	3,35E+23	4,53E+57	1,1E+77
a_{146}	108638,4855	3,95E+11	4,91E+23	1,15E+58	3,82E+77
a_{147}	117867,6403	4,76E+11	7,2E+23	2,93E+58	1,33E+78
a_{148}	127880,8386	5,75E+11	1,06E+24	7,47E+58	4,62E+78
a_{149}	138744,6871	6,93E+11	1,55E+24	1,9E+59	1,61E+79
a_{150}	150531,4511	8,36E+11	2,27E+24	4,83E+59	5,59E+79
a_{151}	163319,535	1,01E+12	3,32E+24	1,23E+60	1,95E+80
a_{152}	177194,0037	1,22E+12	4,87E+24	3,13E+60	6,77E+80
a_{153}	192247,1488	1,47E+12	7,14E+24	7,96E+60	2,36E+81
a_{154}	208579,1023	1,77E+12	1,05E+25	2,03E+61	8,19E+81
a_{155}	226298,5027	2,13E+12	1,53E+25	5,16E+61	2,85E+82
a_{156}	245523,2176	2,57E+12	2,25E+25	1,31E+62	9,92E+82
a_{157}	266381,1278	3,11E+12	3,29E+25	3,34E+62	3,45E+83
a_{158}	289010,9781	3,75E+12	4,82E+25	8,49E+62	1,2E+84
a_{159}	313563,2998	4,52E+12	7,07E+25	2,16E+63	4,18E+84
a_{160}	340201,4126	5,45E+12	1,04E+26	5,5E+63	1,45E+85
a_{161}	369102,5104	6,57E+12	1,52E+26	1,4E+64	5,05E+85
a_{162}	400458,8404	7,93E+12	2,23E+26	3,56E+64	1,76E+86
a_{163}	434478,9816	9,56E+12	3,26E+26	9,06E+64	6,12E+86
a_{164}	471389,2327	1,15E+13	4,78E+26	2,31E+65	2,13E+87
a_{165}	511435,1167	1,39E+13	7,01E+26	5,87E+65	7,4E+87
a_{166}	554883,0149	1,68E+13	1,03E+27	1,49E+66	2,57E+88
a_{167}	602021,9382	2,02E+13	1,5E+27	3,8E+66	8,96E+88
a_{168}	653165,4498	2,44E+13	2,21E+27	9,66E+66	3,12E+89

Solving equations - using modified Fibonacci sequences 42
- an observation

a_{169}	708653,7513	2,95E+13	3,23E+27	2,46E+67	1,08E+90
a_{170}	768855,9451	3,55E+13	4,74E+27	6,26E+67	3,77E+90
a_{171}	834172,4901	4,29E+13	6,94E+27	1,59E+68	1,31E+91
a_{172}	905037,8652	5,17E+13	1,02E+28	4,05E+68	4,56E+91
a_{173}	981923,4598	6,24E+13	1,49E+28	1,03E+69	1,59E+92
a_{174}	1065340,709	7,52E+13	2,19E+28	2,62E+69	5,52E+92
a_{175}	1155844,495	9,07E+13	3,2E+28	6,67E+69	1,92E+93
a_{176}	1254036,841	1,09E+14	4,69E+28	1,7E+70	6,68E+93
a_{177}	1360570,912	1,32E+14	6,88E+28	4,32E+70	2,33E+94
a_{178}	1476155,362	1,59E+14	1,01E+29	1,1E+71	8,09E+94
a_{179}	1601559,046	1,92E+14	1,48E+29	2,8E+71	2,81E+95
a_{180}	1737616,137	2,32E+14	2,17E+29	7,12E+71	9,79E+95
a_{181}	1885231,673	2,79E+14	3,17E+29	1,81E+72	3,41E+96
a_{182}	2045387,578	3,37E+14	4,65E+29	4,61E+72	1,18E+97
a_{183}	2219149,191	4,06E+14	6,82E+29	1,17E+73	4,12E+97
a_{184}	2407672,359	4,9E+14	9,99E+29	2,98E+73	1,43E+98
a_{185}	2612211,117	5,91E+14	1,46E+30	7,59E+73	4,99E+98
a_{186}	2834126,036	7,13E+14	2,15E+30	1,93E+74	1,7E+99
a_{187}	3074893,272	8,6E+14	3,15E+30	4,92E+74	6E+99
a_{188}	3336114,383	1,04E+15	4,61E+30	1,25E+75	2,1E+100
a_{189}	3619526,987	1,25E+15	6,76E+30	3,18E+75	7,3E+100
a_{190}	3927016,314	1,51E+15	9,9E+30	8,1E+75	2,5E+101
a_{191}	4260627,752	1,82E+15	1,45E+31	2,06E+76	8,8E+101
a_{192}	4622580,451	2,2E+15	2,13E+31	5,24E+76	3,1E+102
a_{193}	5015282,082	2,65E+15	3,12E+31	1,33E+77	1,1E+103
a_{194}	5441344,858	3,2E+15	4,57E+31	3,4E+77	3,7E+103
a_{195}	5903602,903	3,86E+15	6,69E+31	8,64E+77	1,3E+104
a_{196}	6405131,111	4,65E+15	9,81E+31	2,2E+78	4,5E+104
a_{197}	6949265,597	5,61E+15	1,44E+32	5,59E+78	1,6E+105
a_{198}	7539625,887	6,77E+15	2,11E+32	1,42E+79	5,5E+105
a_{199}	8180138,998	8,16E+15	3,09E+32	3,62E+79	1,9E+106
a_{200}	8875065,558	9,84E+15	4,53E+32	9,22E+79	6,6E+106

1.3 Calculation example for v-1=3

$$\frac{a_{200}}{a_{199}}$$

	1,079494742	1,18177	1,380278	2,09209	2,636661
b	0,1	0,3	1	10	30
a_1	0	0	0	0	0
a_2	1	1	1	1	1
a_3	1	1	1	1	1
a_4	1	1	1	1	1
a_5	1	1	1	1	1
a_6	1,1	1,3	2	11	31
a_7	1,2	1,6	3	21	61
a_8	1,3	1,9	4	31	91
a_9	1,4	2,2	5	41	121
a_{10}	1,51	2,59	7	151	1051
a_{11}	1,63	3,07	10	361	2881
a_{12}	1,76	3,64	14	671	5611
a_{13}	1,9	4,3	19	1081	9241
a_{14}	2,051	5,077	26	2591	40771
a_{15}	2,214	5,998	36	6201	127201
a_{16}	2,39	7,09	50	12911	295531
a_{17}	2,58	8,38	69	23721	572761
a_{18}	2,7851	9,9031	95	49631	1795891
a_{19}	3,0065	11,7025	131	111641	5611921
a_{20}	3,2455	13,8295	181	240751	14477851
a_{21}	3,5035	16,3435	250	477961	31660681
a_{22}	3,78201	19,31443	345	974271	85537411
a_{23}	4,08266	22,82518	476	2090681	2,54E+08
a_{24}	4,40721	26,97403	657	4498191	6,88E+08
a_{25}	4,75756	31,87708	907	9277801	1,64E+09
a_{26}	5,135761	37,67141	1252	19020511	4,2E+09
a_{27}	5,544027	44,51896	1728	39927321	1,18E+10
a_{28}	5,984748	52,61117	2385	84909231	3,25E+10

a_{29}	6,460504	62,1743	3292	1,78E+08	8,16E+10
a_{30}	6,9740801	73,47572	4544	3,68E+08	2,08E+11
a_{31}	7,5284828	86,83141	6272	7,67E+08	5,62E+11
a_{32}	8,1269576	102,6148	8657	1,62E+09	1,54E+12
a_{33}	8,773008	121,267	11949	3,39E+09	3,98E+12
a_{34}	9,47041601	143,3098	16493	7,07E+09	1,02E+13
a_{35}	10,22326429	169,3592	22765	1,47E+10	2,71E+13
a_{36}	11,03596005	200,1436	31422	3,09E+10	7,32E+13
a_{37}	11,91326085	236,5237	43371	6,48E+10	1,93E+14
a_{38}	12,86030245	279,5167	59864	1,36E+11	4,99E+14
a_{39}	13,88262888	330,3244	82629	2,83E+11	1,31E+15
a_{40}	14,98622489	390,3675	114051	5,92E+11	3,51E+15
a_{41}	16,17755097	461,3246	157422	1,24E+12	9,29E+15
a_{42}	17,46358122	545,1796	217286	2,6E+12	2,43E+16
a_{43}	18,8518441	644,2769	299915	5,43E+12	6,36E+16
a_{44}	20,35046659	761,3872	413966	1,13E+13	1,69E+17
a_{45}	21,96822169	899,7846	571388	2,38E+13	4,48E+17
a_{46}	23,71457981	1063,338	788674	4,97E+13	1,18E+18
a_{47}	25,59976422	1256,622	1088589	1,04E+14	3,08E+18
a_{48}	27,63481088	1485,038	1502555	2,17E+14	8,15E+18
a_{49}	29,83163305	1754,973	2073943	4,55E+14	2,16E+19
a_{50}	32,20309103	2073,975	2862617	9,52E+14	5,68E+19
a_{51}	34,76306745	2450,961	3951206	1,99E+15	1,49E+20
a_{52}	37,52654854	2896,472	5453761	4,17E+15	3,94E+20
a_{53}	40,50971184	3422,964	7527704	8,72E+15	1,04E+21
a_{54}	43,73002095	4045,157	10390321	1,82E+16	2,75E+21
a_{55}	47,20632769	4780,445	14341527	3,82E+16	7,23E+21
a_{56}	50,95898255	5649,387	19795288	7,98E+16	1,9E+22
a_{57}	55,00995373	6676,276	27322992	1,67E+17	5,03E+22
a_{58}	59,38295583	7889,823	37713313	3,49E+17	1,33E+23
a_{59}	64,10358859	9323,956	52054840	7,31E+17	3,49E+23
a_{60}	69,19948685	11018,77	71850128	1,53E+18	9,21E+23
a_{61}	74,70048222	13021,66	99173120	3,2E+18	2,43E+24
a_{62}	80,6387778	15388,6	1,37E+08	6,69E+18	6,41E+24
a_{63}	87,04913666	18185,79	1,89E+08	1,4E+19	1,69E+25
a_{64}	93,96908535	21491,42	2,61E+08	2,93E+19	4,45E+25

a_{65}	101,4391336	25397,92	3,6E+08	6,13E+19	1,17E+26
a_{66}	109,5030114	30014,5	4,97E+08	1,28E+20	3,1E+26
a_{67}	118,207925	35470,23	6,86E+08	2,68E+20	8,16E+26
a_{68}	127,6048336	41917,66	9,47E+08	5,61E+20	2,15E+27
a_{69}	137,7487469	49537,04	1,31E+09	1,17E+21	5,67E+27
a_{70}	148,699048	58541,39	1,8E+09	2,46E+21	1,5E+28
a_{71}	160,5198405	69182,46	2,49E+09	5,14E+21	3,95E+28
a_{72}	173,2803239	81757,75	3,44E+09	1,07E+22	1,04E+29
a_{73}	187,0551986	96618,87	4,74E+09	2,25E+22	2,74E+29
a_{74}	201,9251034	114181,3	6,55E+09	4,7E+22	7,23E+29
a_{75}	217,9770875	134936	9,03E+09	9,84E+22	1,91E+30
a_{76}	235,3051198	159463,3	1,25E+10	2,06E+23	5,03E+30
a_{77}	254,0106397	188449	1,72E+10	4,31E+23	1,33E+31
a_{78}	274,20315	222703,4	2,38E+10	9,01E+23	3,49E+31
a_{79}	296,0008588	263184,2	3,28E+10	1,89E+24	9,22E+31
a_{80}	319,5313708	311023,2	4,53E+10	3,94E+24	2,43E+32
a_{81}	344,9324347	367557,9	6,25E+10	8,25E+24	6,41E+32
a_{82}	372,3527497	434368,9	8,62E+10	1,73E+25	1,69E+33
a_{83}	401,9528356	513324,2	1,19E+11	3,61E+25	4,45E+33
a_{84}	433,9059727	606631,1	1,64E+11	7,56E+25	1,17E+34
a_{85}	468,3992162	716898,5	2,27E+11	1,58E+26	3,1E+34
a_{86}	505,6344912	847209,2	3,13E+11	3,31E+26	8,16E+34
a_{87}	545,8297747	1001206	4,32E+11	6,92E+26	2,15E+35
a_{88}	589,220372	1183196	5,96E+11	1,45E+27	5,68E+35
a_{89}	636,0602936	1398265	8,23E+11	3,03E+27	1,5E+36
a_{90}	686,6237427	1652428	1,14E+12	6,34E+27	3,95E+36
a_{91}	741,2067202	1952790	1,57E+12	1,33E+28	1,04E+37
a_{92}	800,1287574	2307749	2,16E+12	2,77E+28	2,74E+37
a_{93}	863,7347867	2727228	2,99E+12	5,8E+28	7,23E+37
a_{94}	932,397161	3222957	4,12E+12	1,21E+29	1,91E+38
a_{95}	1006,517833	3808794	5,69E+12	2,54E+29	5,03E+38
a_{96}	1086,530709	4501118	7,86E+12	5,31E+29	1,33E+39
a_{97}	1172,904187	5319287	1,08E+13	1,11E+30	3,5E+39
a_{98}	1266,143904	6286174	1,5E+13	2,33E+30	9,22E+39
a_{99}	1366,795687	7428812	2,07E+13	4,86E+30	2,43E+40
a_{100}	1475,448758	8779148	2,85E+13	1,02E+31	6,41E+40

Solving equations - using modified Fibonacci sequences
- an observation

a_{101}	1592,739176	10374934	3,94E+13	2,13E+31	1,69E+41
a_{102}	1719,353567	12260786	5,43E+13	4,45E+31	4,45E+41
a_{103}	1856,033136	14489429	7,5E+13	9,32E+31	1,17E+42
a_{104}	2003,578011	17123174	1,04E+14	1,95E+32	3,1E+42
a_{105}	2162,851929	20235654	1,43E+14	4,08E+32	8,16E+42
a_{106}	2334,787286	23913889	1,97E+14	8,53E+32	2,15E+43
a_{107}	2520,390599	28260718	2,72E+14	1,79E+33	5,68E+43
a_{108}	2720,7484	33397670	3,76E+14	3,73E+33	1,5E+44
a_{109}	2937,033593	39468366	5,19E+14	7,81E+33	3,95E+44
a_{110}	3170,512322	46642533	7,16E+14	1,63E+34	1,04E+45
a_{111}	3422,551382	55120749	9,88E+14	3,42E+34	2,74E+45
a_{112}	3694,626222	65140050	1,36E+15	7,15E+34	7,23E+45
a_{113}	3988,329581	76980560	1,88E+15	1,5E+35	1,91E+46
a_{114}	4305,380813	90973320	2,6E+15	3,13E+35	5,03E+46
a_{115}	4647,635951	1,08E+08	3,59E+15	6,55E+35	1,33E+47
a_{116}	5017,098574	1,27E+08	4,95E+15	1,37E+36	3,5E+47
a_{117}	5415,931532	1,5E+08	6,83E+15	2,87E+36	9,22E+47
a_{118}	5846,469613	1,77E+08	9,43E+15	6E+36	2,43E+48
a_{119}	6311,233208	2,1E+08	1,3E+16	1,26E+37	6,41E+48
a_{120}	6812,943065	2,48E+08	1,8E+16	2,63E+37	1,69E+49
a_{121}	7354,536219	2,93E+08	2,48E+16	5,49E+37	4,45E+49
a_{122}	7939,18318	3,46E+08	3,42E+16	1,15E+38	1,17E+50
a_{123}	8570,306501	4,09E+08	4,72E+16	2,4E+38	3,1E+50
a_{124}	9251,600807	4,83E+08	6,52E+16	5,03E+38	8,16E+50
a_{125}	9987,054429	5,71E+08	9E+16	1,05E+39	2,15E+51
a_{126}	10780,97275	6,75E+08	1,24E+17	2,2E+39	5,68E+51
a_{127}	11638,0034	7,98E+08	1,71E+17	4,61E+39	1,5E+52
a_{128}	12563,16348	9,43E+08	2,37E+17	9,64E+39	3,95E+52
a_{129}	13561,86892	1,11E+09	3,27E+17	2,02E+40	1,04E+53
a_{130}	14639,9662	1,32E+09	4,51E+17	4,22E+40	2,74E+53
a_{131}	15803,76654	1,56E+09	6,22E+17	8,82E+40	7,23E+53
a_{132}	17060,08288	1,84E+09	8,59E+17	1,85E+41	1,91E+54
a_{133}	18416,26978	2,17E+09	1,19E+18	3,86E+41	5,03E+54
a_{134}	19880,26639	2,57E+09	1,64E+18	8,08E+41	1,33E+55
a_{135}	21460,64305	3,03E+09	2,26E+18	1,69E+42	3,5E+55
a_{136}	23166,65134	3,59E+09	3,12E+18	3,54E+42	9,22E+55

Solving equations - using modified Fibonacci sequences   47
- an observation

a_{137}	25008,27831	4,24E+09	4,3E+18	7,4E+42	2,43E+56
a_{138}	26996,30495	5,01E+09	5,94E+18	1,55E+43	6,41E+56
a_{139}	29142,36926	5,92E+09	8,2E+18	3,24E+43	1,69E+57
a_{140}	31459,03439	6,99E+09	1,13E+19	6,77E+43	4,45E+57
a_{141}	33959,86222	8,27E+09	1,56E+19	1,42E+44	1,17E+58
a_{142}	36659,49272	9,77E+09	2,16E+19	2,96E+44	3,1E+58
a_{143}	39573,72964	1,15E+10	2,98E+19	6,2E+44	8,16E+58
a_{144}	42719,63308	1,36E+10	4,11E+19	1,3E+45	2,15E+59
a_{145}	46115,61931	1,61E+10	5,67E+19	2,71E+45	5,68E+59
a_{146}	49781,56858	1,91E+10	7,83E+19	5,68E+45	1,5E+60
a_{147}	53738,94154	2,25E+10	1,08E+20	1,19E+46	3,95E+60
a_{148}	58010,90485	2,66E+10	1,49E+20	2,49E+46	1,04E+61
a_{149}	62622,46678	3,14E+10	2,06E+20	5,2E+46	2,74E+61
a_{150}	67600,62364	3,72E+10	2,84E+20	1,09E+47	7,23E+61
a_{151}	72974,51779	4,39E+10	3,92E+20	2,28E+47	1,91E+62
a_{152}	78775,60828	5,19E+10	5,41E+20	4,76E+47	5,03E+62
a_{153}	85037,85496	6,13E+10	7,47E+20	9,96E+47	1,33E+63
a_{154}	91797,91732	7,25E+10	1,03E+21	2,08E+48	3,5E+63
a_{155}	99095,3691	8,57E+10	1,42E+21	4,36E+48	9,22E+63
a_{156}	106972,9299	1,01E+11	1,96E+21	9,12E+48	2,43E+64
a_{157}	115476,7154	1,2E+11	2,71E+21	1,91E+49	6,41E+64
a_{158}	124656,5072	1,41E+11	3,74E+21	3,99E+49	1,69E+65
a_{159}	134566,0441	1,67E+11	5,17E+21	8,35E+49	4,45E+65
a_{160}	145263,3371	1,97E+11	7,13E+21	1,75E+50	1,17E+66
a_{161}	156811,0086	2,33E+11	9,84E+21	3,66E+50	3,1E+66
a_{162}	169276,6593	2,76E+11	1,36E+22	7,65E+50	8,17E+66
a_{163}	182733,2637	3,26E+11	1,87E+22	1,6E+51	2,15E+67
a_{164}	197259,5974	3,85E+11	2,59E+22	3,35E+51	5,68E+67
a_{165}	212940,6983	4,55E+11	3,57E+22	7E+51	1,5E+68
a_{166}	229868,3642	5,38E+11	4,93E+22	1,47E+52	3,95E+68
a_{167}	248141,6906	6,36E+11	6,81E+22	3,07E+52	1,04E+69
a_{168}	267867,6503	7,51E+11	9,39E+22	6,41E+52	2,74E+69
a_{169}	289161,7202	8,88E+11	1,3E+23	1,34E+53	7,23E+69
a_{170}	312148,5566	1,05E+12	1,79E+23	2,81E+53	1,91E+70
a_{171}	336962,7256	1,24E+12	2,47E+23	5,87E+53	5,03E+70
a_{172}	363749,4907	1,46E+12	3,41E+23	1,23E+54	1,33E+71

Solving equations - using modified Fibonacci sequences   48
- an observation

a_{173}	392665,6627	1,73E+12	4,71E+23	2,57E+54	3,5E+71
a_{174}	423880,5184	2,05E+12	6,5E+23	5,38E+54	9,22E+71
a_{175}	457576,7909	2,42E+12	8,97E+23	1,13E+55	2,43E+72
a_{176}	493951,74	2,86E+12	1,24E+24	2,35E+55	6,41E+72
a_{177}	533218,3063	3,38E+12	1,71E+24	4,92E+55	1,69E+73
a_{178}	575606,3581	3,99E+12	2,36E+24	1,03E+56	4,45E+73
a_{179}	621364,0372	4,72E+12	3,25E+24	2,16E+56	1,17E+74
a_{180}	670759,2112	5,57E+12	4,49E+24	4,51E+56	3,1E+74
a_{181}	724081,0418	6,59E+12	6,2E+24	9,43E+56	8,17E+74
a_{182}	781641,6776	7,78E+12	8,56E+24	1,97E+57	2,15E+75
a_{183}	843778,0813	9,2E+12	1,18E+25	4,13E+57	5,68E+75
a_{184}	910854,0025	1,09E+13	1,63E+25	8,64E+57	1,5E+76
a_{185}	983262,1066	1,28E+13	2,25E+25	1,81E+58	3,95E+76
a_{186}	1061426,274	1,52E+13	3,11E+25	3,78E+58	1,04E+77
a_{187}	1145804,083	1,79E+13	4,29E+25	7,91E+58	2,74E+77
a_{188}	1236889,483	2,12E+13	5,92E+25	1,65E+59	7,23E+77
a_{189}	1335215,693	2,51E+13	8,17E+25	3,46E+59	1,91E+78
a_{190}	1441358,321	2,96E+13	1,13E+26	7,24E+59	5,03E+78
a_{191}	1555938,729	3,5E+13	1,56E+26	1,52E+60	1,33E+79
a_{192}	1679627,677	4,14E+13	2,15E+26	3,17E+60	3,5E+79
a_{193}	1813149,247	4,89E+13	2,96E+26	6,63E+60	9,22E+79
a_{194}	1957285,079	5,78E+13	4,09E+26	1,39E+61	2,43E+80
a_{195}	2112878,952	6,82E+13	5,65E+26	2,9E+61	6,41E+80
a_{196}	2280841,72	8,07E+13	7,8E+26	6,07E+61	1,69E+81
a_{197}	2462156,644	9,53E+13	1,08E+27	1,27E+62	4,46E+81
a_{198}	2657885,152	1,13E+14	1,49E+27	2,66E+62	1,17E+82
a_{199}	2869173,047	1,33E+14	2,05E+27	5,56E+62	3,1E+82
a_{200}	3097257,219	1,57E+14	2,83E+27	1,16E+63	8,17E+82

1.4 Calculation example for v-1=10

$$\cdot \ \frac{a_{200}}{a_{199}}$$

a_{199}	1,057288283	1,107786	1,184276	1,385008	1,504925
b	0,1	0,3	1	10	30
a_1	0	0	0	0	0
a_2	1	1	1	1	1
a_3	1	1	1	1	1
a_4	1	1	1	1	1
a_5	1	1	1	1	1
a_6	1	1	1	1	1
a_7	1	1	1	1	1
a_8	1	1	1	1	1
a_9	1	1	1	1	1
a_{10}	1	1	1	1	1
a_{11}	1	1	1	1	1
a_{12}	1	1	1	1	1
a_{13}	1,1	1,3	2	11	31
a_{14}	1,2	1,6	3	21	61
a_{15}	1,3	1,9	4	31	91
a_{16}	1,4	2,2	5	41	121
a_{17}	1,5	2,5	6	51	151
a_{18}	1,6	2,8	7	61	181
a_{19}	1,7	3,1	8	71	211
a_{20}	1,8	3,4	9	81	241
a_{21}	1,9	3,7	10	91	271
a_{22}	2	4	11	101	301
a_{23}	2,1	4,3	12	111	331
a_{24}	2,21	4,69	14	221	1261
a_{25}	2,33	5,17	17	431	3091
a_{26}	2,46	5,74	21	741	5821
a_{27}	2,6	6,4	26	1151	9451
a_{28}	2,75	7,15	32	1661	13981

a_{29}	2,91	7,99	39	2271	19411
a_{30}	3,08	8,92	47	2981	25741
a_{31}	3,26	9,94	56	3791	32971
a_{32}	3,45	11,05	66	4701	41101
a_{33}	3,65	12,25	77	5711	50131
a_{34}	3,86	13,54	89	6821	60061
a_{35}	4,081	14,947	103	9031	97891
a_{36}	4,314	16,498	120	13341	190621
a_{37}	4,56	18,22	141	20751	365251
a_{38}	4,82	20,14	167	32261	648781
a_{39}	5,095	22,285	199	48871	1068211
a_{40}	5,386	24,682	238	71581	1650541
a_{41}	5,694	27,358	285	101391	2422771
a_{42}	6,02	30,34	341	139301	3411901
a_{43}	6,365	33,655	407	186311	4644931
a_{44}	6,73	37,33	484	243421	6148861
a_{45}	7,116	41,392	573	311631	7950691
a_{46}	7,5241	45,8761	676	401941	10887421
a_{47}	7,9555	50,8255	796	535351	16606051
a_{48}	8,4115	56,2915	937	742861	27563581
a_{49}	8,8935	62,3335	1104	1065471	47027011
a_{50}	9,403	69,019	1303	1554181	79073341
a_{51}	9,9416	76,4236	1541	2269991	1,29E+08
a_{52}	10,511	84,631	1826	3283901	2,01E+08
a_{53}	11,113	93,733	2167	4676911	3,04E+08
a_{54}	11,7495	103,8295	2574	6540021	4,43E+08
a_{55}	12,4225	115,0285	3058	8974231	6,27E+08
a_{56}	13,1341	127,4461	3631	12090541	8,66E+08
a_{57}	13,88651	141,2089	4307	16109951	1,19E+09
a_{58}	14,68206	156,4566	5103	21463461	1,69E+09
a_{59}	15,52321	173,344	6040	28892071	2,52E+09
a_{60}	16,41256	192,0441	7144	39546781	3,93E+09
a_{61}	17,35286	212,7498	8447	55088591	6,3E+09
a_{62}	18,34702	235,6769	9988	77788501	1,02E+10
a_{63}	19,39812	261,0662	11814	1,11E+08	1,62E+10
a_{64}	20,50942	289,1861	13981	1,57E+08	2,53E+10

a_{65}	21,68437	320,3349	16555	2,23E+08	3,86E+10
a_{66}	22,92662	354,8435	19613	3,13E+08	5,74E+10
a_{67}	24,24003	393,0773	23244	4,33E+08	8,34E+10
a_{68}	25,628681	435,44	27551	5,95E+08	1,19E+11
a_{69}	27,096887	482,3769	32654	8,09E+08	1,7E+11
a_{70}	28,649208	534,3802	38694	1,1E+09	2,45E+11
a_{71}	30,290464	591,9934	45838	1,49E+09	3,63E+11
a_{72}	32,02575	655,8183	54285	2,04E+09	5,52E+11
a_{73}	33,860452	726,5214	64273	2,82E+09	8,57E+11
a_{74}	35,800264	804,8412	76087	3,93E+09	1,34E+12
a_{75}	37,851206	891,597	90068	5,5E+09	2,1E+12
a_{76}	40,019643	987,6975	106623	7,73E+09	3,26E+12
a_{77}	42,312305	1094,151	126236	1,09E+10	4,98E+12
a_{78}	44,736308	1212,074	149480	1,52E+10	7,48E+12
a_{79}	47,2991761	1342,706	177031	2,11E+10	1,11E+13
a_{80}	50,0088648	1487,419	209685	2,92E+10	1,62E+13
a_{81}	52,8737856	1647,733	248379	4,02E+10	2,35E+13
a_{82}	55,902832	1825,331	294217	5,51E+10	3,44E+13
a_{83}	59,105407	2022,076	348502	7,56E+10	5,1E+13
a_{84}	62,4914522	2240,033	412775	1,04E+11	7,67E+13
a_{85}	66,0714786	2481,485	488862	1,43E+11	1,17E+14
a_{86}	69,8565992	2748,964	578930	1,98E+11	1,8E+14
a_{87}	73,8585635	3045,273	685553	2,75E+11	2,78E+14
a_{88}	78,089794	3373,519	811789	3,84E+11	4,27E+14
a_{89}	82,5634248	3737,141	961269	5,36E+11	6,52E+14
a_{90}	87,29334241	4139,952	1138300	7,47E+11	9,84E+14
a_{91}	92,29422889	4586,178	1347985	1,04E+12	1,47E+15
a_{92}	97,58160745	5080,498	1596364	1,44E+12	2,17E+15
a_{93}	103,1718907	5628,097	1890581	1,99E+12	3,21E+15
a_{94}	109,0824314	6234,72	2239083	2,75E+12	4,74E+15
a_{95}	115,3315766	6906,73	2651858	3,79E+12	7,04E+15
a_{96}	121,9387244	7651,176	3140720	5,22E+12	1,05E+16
a_{97}	128,9243844	8475,865	3719650	7,2E+12	1,59E+16
a_{98}	136,3102407	9389,447	4405203	9,95E+12	2,43E+16
a_{99}	144,1192201	10401,5	5216992	1,38E+13	3,71E+16
a_{100}	152,3755626	11522,64	6178261	1,92E+13	5,67E+16

Solving equations - using modified Fibonacci sequences
- an observation

a_{101}	161,1048968	12764,63	7316561	2,66E+13	8,62E+16
a_{102}	170,3343197	14140,48	8664546	3,7E+13	1,3E+17
a_{103}	180,0924805	15664,63	10260910	5,14E+13	1,95E+17
a_{104}	190,4096695	17353,06	12151491	7,14E+13	2,92E+17
a_{105}	201,3179127	19223,48	14390574	9,89E+13	4,34E+17
a_{106}	212,8510703	21295,5	17042432	1,37E+14	6,45E+17
a_{107}	225,0449428	23590,85	20183152	1,89E+14	9,61E+17
a_{108}	237,9373812	26133,61	23902802	2,61E+14	1,44E+18
a_{109}	251,5684053	28950,44	28308005	3,6E+14	2,17E+18
a_{110}	265,9803273	32070,89	33524997	4,98E+14	3,28E+18
a_{111}	281,2178835	35527,69	39703258	6,9E+14	4,98E+18
a_{112}	297,3283732	39357,08	47019819	9,56E+14	7,57E+18
a_{113}	314,3618052	43599,22	55684365	1,33E+15	1,15E+19
a_{114}	332,3710532	48298,61	65945275	1,84E+15	1,73E+19
a_{115}	351,4120202	53504,53	78096766	2,55E+15	2,61E+19
a_{116}	371,5438114	59271,57	92487340	3,54E+15	3,91E+19
a_{117}	392,8289185	65660,22	1,1E+08	4,91E+15	5,84E+19
a_{118}	415,3334128	72737,48	1,3E+08	6,8E+15	8,73E+19
a_{119}	439,1271509	80577,56	1,54E+08	9,41E+15	1,3E+20
a_{120}	464,2839914	89262,69	1,82E+08	1,3E+16	1,96E+20
a_{121}	490,8820241	98883,96	2,15E+08	1,8E+16	2,94E+20
a_{122}	519,0038125	109542,3	2,55E+08	2,49E+16	4,43E+20
a_{123}	548,7366498	121349,4	3,02E+08	3,45E+16	6,7E+20
a_{124}	580,1728303	134429,2	3,58E+08	4,77E+16	1,01E+21
a_{125}	613,4099356	148918,7	4,24E+08	6,61E+16	1,53E+21
a_{126}	648,5511377	164970,1	5,02E+08	9,17E+16	2,32E+21
a_{127}	685,7055188	182751,6	5,94E+08	1,27E+17	3,49E+21
a_{128}	724,9884106	202449,6	7,04E+08	1,76E+17	5,24E+21
a_{129}	766,5217519	224270,9	8,34E+08	2,44E+17	7,86E+21
a_{130}	810,434467	248444,2	9,87E+08	3,38E+17	1,18E+22
a_{131}	856,8628661	275223	1,17E+09	4,68E+17	1,76E+22
a_{132}	905,9510686	304888,2	1,38E+09	6,48E+17	2,65E+22
a_{133}	957,8514498	337750,8	1,64E+09	8,97E+17	3,98E+22
a_{134}	1012,725115	374155,6	1,94E+09	1,24E+18	5,99E+22
a_{135}	1070,742398	414484,4	2,3E+09	1,72E+18	9,03E+22
a_{136}	1132,083391	459160	2,72E+09	2,38E+18	1,36E+23

Solving equations - using modified Fibonacci sequences 53
- an observation

a_{137}	1196,938505	508651	3,23E+09	3,3E+18	2,06E+23
a_{138}	1265,509057	563476,5	3,82E+09	4,57E+18	3,11E+23
a_{139}	1338,007898	624211,4	4,52E+09	6,33E+18	4,68E+23
a_{140}	1414,660073	691492,7	5,36E+09	8,77E+18	7,04E+23
a_{141}	1495,70352	766025,9	6,34E+09	1,22E+19	1,06E+24
a_{142}	1581,389807	848592,8	7,51E+09	1,68E+19	1,59E+24
a_{143}	1671,984913	940059,3	8,9E+09	2,33E+19	2,38E+24
a_{144}	1767,770058	1041385	1,05E+10	3,23E+19	3,57E+24
a_{145}	1869,04257	1153631	1,25E+10	4,47E+19	5,37E+24
a_{146}	1976,11681	1277977	1,48E+10	6,19E+19	8,08E+24
a_{147}	2089,325149	1415725	1,75E+10	8,57E+19	1,22E+25
a_{148}	2209,018999	1568320	2,07E+10	1,19E+20	1,83E+25
a_{149}	2335,569905	1737363	2,45E+10	1,64E+20	2,77E+25
a_{150}	2469,370695	1924626	2,91E+10	2,28E+20	4,17E+25
a_{151}	2610,836702	2132074	3,44E+10	3,15E+20	6,28E+25
a_{152}	2760,407054	2361882	4,08E+10	4,37E+20	9,45E+25
a_{153}	2918,546035	2616460	4,83E+10	6,05E+20	1,42E+26
a_{154}	3085,744526	2898477	5,72E+10	8,39E+20	2,14E+26
a_{155}	3262,521532	3210893	6,77E+10	1,16E+21	3,21E+26
a_{156}	3449,425789	3556982	8,02E+10	1,61E+21	4,82E+26
a_{157}	3647,03747	3940375	9,5E+10	2,23E+21	7,24E+26
a_{158}	3855,969985	4365092	1,12E+11	3,08E+21	1,09E+27
a_{159}	4076,871885	4835588	1,33E+11	4,27E+21	1,64E+27
a_{160}	4310,428875	5356797	1,58E+11	5,92E+21	2,47E+27
a_{161}	4557,365945	5934185	1,87E+11	8,19E+21	3,72E+27
a_{162}	4818,449615	6573807	2,21E+11	1,13E+22	5,6E+27
a_{163}	5094,49032	7282372	2,62E+11	1,57E+22	8,44E+27
a_{164}	5386,344924	8067310	3,1E+11	2,18E+22	1,27E+28
a_{165}	5694,919377	8936853	3,68E+11	3,02E+22	1,91E+28
a_{166}	6021,17153	9900121	4,35E+11	4,18E+22	2,87E+28
a_{167}	6366,114109	10967215	5,15E+11	5,79E+22	4,32E+28
a_{168}	6730,817856	12149328	6,1E+11	8,01E+22	6,49E+28
a_{169}	7116,414854	13458856	7,23E+11	1,11E+23	9,76E+28
a_{170}	7524,102043	14909532	8,56E+11	1,54E+23	1,47E+29
a_{171}	7955,14493	16516571	1,01E+12	2,13E+23	2,21E+29
a_{172}	8410,881525	18296827	1,2E+12	2,95E+23	3,32E+29

Solving equations - using modified Fibonacci sequences 54
- an observation

a_{173}	8892,726486	20268969	1,42E+12	4,08E+23	5,01E+29
a_{174}	9402,175518	22453681	1,68E+12	5,65E+23	7,54E+29
a_{175}	9940,810011	24873874	1,99E+12	7,83E+23	1,13E+30
a_{176}	10510,30195	27554930	2,36E+12	1,08E+24	1,71E+30
a_{177}	11112,4191	30524966	2,8E+12	1,5E+24	2,57E+30
a_{178}	11749,03051	33815130	3,31E+12	2,08E+24	3,87E+30
a_{179}	12422,1123	37459929	3,92E+12	2,88E+24	5,81E+30
a_{180}	13133,75378	41497586	4,65E+12	3,99E+24	8,74E+30
a_{181}	13886,16399	45970445	5,5E+12	5,53E+24	1,31E+31
a_{182}	14681,67848	50925417	6,52E+12	7,66E+24	1,98E+31
a_{183}	15522,76663	56414465	7,72E+12	1,06E+25	2,97E+31
a_{184}	16412,03928	62495155	9,14E+12	1,47E+25	4,48E+31
a_{185}	17352,25683	69231260	1,08E+13	2,03E+25	6,74E+31
a_{186}	18346,33783	76693422	1,28E+13	2,82E+25	1,01E+32
a_{187}	19397,36803	84959901	1,52E+13	3,9E+25	1,53E+32
a_{188}	20508,60994	94117390	1,8E+13	5,4E+25	2,3E+32
a_{189}	21683,51299	1,04E+08	2,13E+13	7,48E+25	3,46E+32
a_{190}	22925,72422	1,15E+08	2,52E+13	1,04E+26	5,2E+32
a_{191}	24239,0996	1,28E+08	2,99E+13	1,44E+26	7,82E+32
a_{192}	25627,716	1,42E+08	3,54E+13	1,99E+26	1,18E+33
a_{193}	27095,88385	1,57E+08	4,19E+13	2,75E+26	1,77E+33
a_{194}	28648,16051	1,74E+08	4,96E+13	3,82E+26	2,66E+33
a_{195}	30289,36444	1,93E+08	5,87E+13	5,28E+26	4E+33
a_{196}	32024,59012	2,13E+08	6,96E+13	7,32E+26	6,03E+33
a_{197}	33859,2239	2,36E+08	8,24E+13	1,01E+27	9,07E+33
a_{198}	35798,96071	2,62E+08	9,76E+13	1,4E+27	1,36E+34
a_{199}	37849,8217	2,9E+08	1,16E+14	1,94E+27	2,05E+34
a_{200}	40018,173	3,21E+08	1,37E+14	2,69E+27	3,09E+34

Annex 2 Example for the main thesis

This Annex lists the calculation results for

$a_n = b_2 a_{n-2} + b_4 a_{n-4} + b_7$ used to determine a root for

$x^7 - b_2 x^5 - b_4 x^3 - b_7 = 0$ in order to support the plausibility check of the main thesis.

Parameters: $b_2 = 0,3; b_4 = 2,5; b_7 = 6$

	a_n	$\dfrac{a_n}{a_{n-1}}$		a_n	$\dfrac{a_n}{a_{n-1}}$
a_1	0		a_{18}	182,826154	1,213898
a_2	1		a_{19}	275,126734	1,504854
a_3	1		a_{20}	499,2302962	1,814547
a_4	1		a_{21}	663,6085702	1,329263
a_5	1		a_{22}	931,1500739	1,403162
a_6	1		a_{23}	1502,092486	1,613158
a_7	1		a_{24}	2431,085443	1,618466
a_8	2,8		a_{25}	3206,606095	1,319002
a_9	8,8	3,142857	a_{26}	4707,961221	1,468207
a_{10}	9,34	1,061364	a_{27}	7712,594821	1,638203
a_{11}	11,14	1,192719	a_{28}	11471,75339	1,487405
a_{12}	15,802	1,418492	a_{29}	15917,19413	1,387512
a_{13}	31,342	1,98342	a_{30}	24223,98399	1,521875
a_{14}	34,0906	1,087697	a_{31}	38643,15795	1,595244
a_{15}	54,0526	1,585557	a_{32}	55186,21525	1,428098
a_{16}	102,53218	1,896896	a_{33}	79633,70003	1,443
a_{17}	150,61078	1,468912	a_{34}	123391,3935	1,549487

		$\dfrac{a_n}{a_{n-1}}$			$\dfrac{a_n}{a_{n-1}}$
a_{35}	189328,5252	1,534374	a_{62}	9202919409	1,486411
a_{36}	270486,1209	1,42866	a_{63}	13767448213	1,495987
a_{37}	401226,7116	1,483354	a_{64}	20627370758	1,498271
a_{38}	621483,2676	1,548958	a_{65}	30721676218	1,489365
a_{39}	924806,6181	1,488064	a_{66}	45776319898	1,490033
a_{40}	1340462,483	1,449452	a_{67}	68534705226	1,497165
a_{41}	2020857,125	1,507582	a_{68}	1,0245E+11	1,494856
a_{42}	3091818,065	1,529954	a_{69}	1,52582E+11	1,489339
a_{43}	4541190,408	1,468777	a_{70}	2,2778E+11	1,492838
a_{44}	6686061,896	1,472315	a_{71}	3,40876E+11	1,49651
a_{45}	10143399,54	1,517096	a_{72}	5,08788E+11	1,492591
a_{46}	15284203,44	1,506813	a_{73}	7,58376E+11	1,490554
a_{47}	22438770,78	1,468102	a_{74}	1,1333E+12	1,494372
a_{48}	33425558,52	1,489634	a_{75}	1,6944E+12	1,495108
a_{49}	50641038,48	1,515039	a_{76}	2,52745E+12	1,491651
a_{50}	75485318,61	1,490596	a_{77}	3,77094E+12	1,491994
a_{51}	111405609,9	1,475858	a_{78}	5,63673E+12	1,49478
a_{52}	167069889,1	1,499654	a_{79}	8,42001E+12	1,493776
a_{53}	251729499,8	1,506732	a_{80}	1,25599E+13	1,491673
a_{54}	373466888	1,483604	a_{81}	1,87531E+13	1,493095
a_{55}	554586225,8	1,484968	a_{82}	2,80262E+13	1,49448
a_{56}	833561020,1	1,503032	a_{83}	4,18407E+13	1,492913
a_{57}	1248611529	1,497925	a_{84}	6,24333E+13	1,492167
a_{58}	1852169185	1,483383	a_{85}	9,32554E+13	1,493681
a_{59}	2763468358	1,492017	a_{86}	1,39315E+14	1,493914
a_{60}	4149930305	1,501711	a_{87}	2,07938E+14	1,492567
a_{61}	6191370657	1,491922	a_{88}	3,10397E+14	1,492738

Solving equations - using modified Fibonacci sequences
- an observation

	$\dfrac{a_n}{a_{n-1}}$			$\dfrac{a_n}{a_{n-1}}$	
a_{89}	4,63677E+14	1,493821	a_{116}	2,32936E+19	1,493161
a_{90}	6,92452E+14	1,493393	a_{117}	3,47804E+19	1,493126
a_{91}	1,03355E+15	1,492591	a_{118}	5,19347E+19	1,493219
a_{92}	1,54326E+15	1,493168	a_{119}	7,75501E+19	1,493224
a_{93}	2,30515E+15	1,493689	a_{120}	1,15794E+20	1,493144
a_{94}	3,44173E+15	1,493063	a_{121}	1,72898E+20	1,493162
a_{95}	5,13779E+15	1,492792	a_{122}	2,58176E+20	1,493224
a_{96}	7,67273E+15	1,493391	a_{123}	3,85507E+20	1,493193
a_{97}	1,14589E+16	1,493461	a_{124}	5,75619E+20	1,493148
a_{98}	1,71074E+16	1,492936	a_{125}	8,59506E+20	1,493187
a_{99}	2,55417E+16	1,493018	a_{126}	1,28343E+21	1,493214
a_{100}	3,8145E+16	1,493438	a_{127}	1,91638E+21	1,493174
a_{101}	5,69602E+16	1,493257	a_{128}	2,86147E+21	1,493162
a_{102}	8,50388E+16	1,492951	a_{129}	4,27274E+21	1,493199
a_{103}	1,26979E+17	1,493185	a_{130}	6,38005E+21	1,493199
a_{104}	1,89628E+17	1,493381	a_{131}	9,52648E+21	1,493168
a_{105}	2,83139E+17	1,493131	a_{132}	1,42247E+22	1,493176
a_{106}	4,22736E+17	1,493033	a_{133}	2,12403E+22	1,4932
a_{107}	6,31258E+17	1,493269	a_{134}	3,17158E+22	1,493187
a_{108}	9,42651E+17	1,493289	a_{135}	4,73571E+22	1,49317
a_{109}	1,40746E+18	1,493084	a_{136}	7,0713E+22	1,493186
a_{110}	2,10151E+18	1,493123	a_{137}	1,05588E+23	1,493196
a_{111}	3,13815E+18	1,493285	a_{138}	1,57662E+23	1,49318
a_{112}	4,68591E+18	1,493209	a_{139}	2,35418E+23	1,493176
a_{113}	6,9965E+18	1,493093	a_{140}	3,51523E+23	1,49319
a_{114}	1,04471E+19	1,493188	a_{141}	5,24891E+23	1,49319
a_{115}	1,56002E+19	1,493261	a_{142}	7,83755E+23	1,493178

		$\dfrac{a_n}{a_{n-1}}$			$\dfrac{a_n}{a_{n-1}}$
a_{143}	1,17029E+24	1,493181	a_{170}	5,87972E+28	1,493185
a_{144}	1,74746E+24	1,493191	a_{171}	8,7795E+28	1,493184
a_{145}	2,60929E+24	1,493185	a_{172}	1,31094E+29	1,493184
a_{146}	3,89613E+24	1,493179	a_{173}	1,95748E+29	1,493185
a_{147}	5,81765E+24	1,493185	a_{174}	2,92287E+29	1,493185
a_{148}	8,68685E+24	1,493189	a_{175}	4,36439E+29	1,493184
a_{149}	1,2971E+25	1,493182	a_{176}	6,51684E+29	1,493184
a_{150}	1,93681E+25	1,493181	a_{177}	9,73084E+29	1,493185
a_{151}	2,89202E+25	1,493187	a_{178}	1,45299E+30	1,493184
a_{152}	4,31833E+25	1,493186	a_{179}	2,16959E+30	1,493184
a_{153}	6,44805E+25	1,493182	a_{180}	3,23959E+30	1,493184
a_{154}	9,62812E+25	1,493183	a_{181}	4,83731E+30	1,493185
a_{155}	1,43766E+26	1,493187	a_{182}	7,223E+30	1,493184
a_{156}	2,14669E+26	1,493185	a_{183}	1,07853E+31	1,493184
a_{157}	3,2054E+26	1,493182	a_{184}	1,61044E+31	1,493184
a_{158}	4,78625E+26	1,493185	a_{185}	2,40468E+31	1,493184
a_{159}	7,14676E+26	1,493186	a_{186}	3,59063E+31	1,493184
a_{160}	1,06714E+27	1,493184	a_{187}	5,36148E+31	1,493184
a_{161}	1,59344E+27	1,493183	a_{188}	8,00567E+31	1,493184
a_{162}	2,3793E+27	1,493185	a_{189}	1,19539E+32	1,493184
a_{163}	3,55273E+27	1,493185	a_{190}	1,78494E+32	1,493184
a_{164}	5,30488E+27	1,493183	a_{191}	2,66525E+32	1,493184
a_{165}	7,92117E+27	1,493184	a_{192}	3,97971E+32	1,493184
a_{166}	1,18278E+28	1,493185	a_{193}	5,94244E+32	1,493184
a_{167}	1,7661E+28	1,493184	a_{194}	8,87316E+32	1,493184
a_{168}	2,63712E+28	1,493183	a_{195}	1,32493E+33	1,493184
a_{169}	3,9377E+28	1,493184	a_{196}	1,97836E+33	1,493184

		$\dfrac{a_n}{a_{n-1}}$			$\dfrac{a_n}{a_{n-1}}$
a_{197}	2,95405E+33	1,493184	a_{200}	9,83465E+33	1,493184
a_{198}	4,41095E+33	1,493184			
a_{199}	6,58636E+33	1,493184			